© 2017
Clement Ampadu
drampadu@hotmail.com

ISBN:978-1-365-92443-9
ID: 20860071
www.lulu.com

All rights reserved. No part of this publication may be produced or transmitted in any form or by any means, electronic or mechanical, including photocopying and recording, or in any information storage and retrieval system, without the prior written permission of the publisher.

Contents

Preface   3

Dedication   4

1 Weak Multiplicative Contraction Mapping Theorem with Multiplicative C-Class Functions in Multiplicative Analogue of Saks Space   5
- 1.1 Brief Summary 5
- 1.2 Preliminaries 5
- 1.3 Main Results 8
- 1.4 Exercises 12
- 1.5 References 13

2 Coincidence and Common Fixed Point Theorems for Implicit f-Weak Multiplicative Contractions in Multiplicative Cone Metric Space   15
- 2.1 Brief Summary 15
- 2.2 Preliminaries 15
- 2.3 Main Results 19
- 2.4 Exercises 22
- 2.5 References 24

3 Fixed Point Theorems for Implicit Generalized Weak Multiplicative Contraction Mappings in Multiplicative Analogue of Modular Spaces   26
- 3.1 Brief Summary 26
- 3.2 Preliminaries 26
- 3.3 Main Results 28
- 3.4 Exercises 31
- 3.5 References 31

4 Fixed Point Theorems for Implicit Weakly Multiplicative Contractions of the Derivative Type in Multiplicative Analogue of T_0-Quasi-Metric Spaces   32
- 4.1 Brief Summary 32
- 4.2 Preliminaries 32
- 4.3 Main Results 36
- 4.4 Exercises 39
- 4.5 References 40

Preface

The notion of C-class function was introduced by A.H. Ansari [A.H. Ansari, Note on $\varphi - \psi$-contractive type mappings and related fixed point, The 2nd Regional Conference on Mathematics and Applications, PNU, September 2014, 377-380]. On the other hand, the notion of multiplicative C-class function was initiated by C.B. Ampadu and then jointly by Ampadu and Ansari [Clement Ampadu and Arslan Hojat Ansari, FIXED POINT THEOREMS IN COMPLETE MULTIPLICATIVE METRIC SPACES WITH APPLICATION TO MULTIPLICATIVE ANALOGUE OF C-CLASS FUNCTIONS, JP Journal of Fixed Point Theory and Applications, August 2016, Volume 11, Issue 2, Pages 113 - 124].

In this monograph we have defined the multiplicative version of weakly contractive mappings [Ya. I. Alber and S. Guerr-Delabriere, Principle of weakly contractive maps Hilbert spaces, New Results in Operator Theory and its Applications (I.Gohberg and Yu. Lyubich, eds.), Oper. Theory Adv. Appl., vol. 98, Birkhauser, Basel, 1997, pp. 7–22] implicitly via the multiplicative C-class function of Ampadu and Ansari [Clement Ampadu and Arslan Hojat Ansari, FIXED POINT THEOREMS IN COMPLETE MULTIPLICATIVE METRIC SPACES WITH APPLICATION TO MULTIPLICATIVE ANALOGUE OF C-CLASS FUNCTIONS, JP Journal of Fixed Point Theory and Applications, August 2016, Volume 11, Issue 2, Pages 113 - 124], and obtained some fixed point theorems for such mappings in the multiplicative analogue of Saks space (Chapter 1), Cone metric space (Chapter 2), Modular space (Chapter 3), and T_0-Quasi-metric space (Chapter 4).

A nice feature of this monograph are the (publishable) exercise set, which begs the reader to explore the beautiful connection between weakly contractive mappings, c-class function, and their multiplicative analogue. The reader will find the notion of multiplicative metric space [A.E. Bashirov, E.M. Kurpnar and A. Ozyapc, Multiplicative calculus and its applications J. Math. Anal. Appl., 337 (2008), 36-48] useful as he or she begins his or her own investigative inquiry.

Prof.Clement Boateng Ampadu

Dedication

Thanking Yahweh, I dedicate this monograph to those who read it, including family, friends, and love ones .

Prof.Clement Boateng Ampadu
May, 2017

Chapter 1

Weak Multiplicative Contraction Mapping Theorem with Multiplicative C-Class Functions in Multiplicative Analogue of Saks Space

1.1 Brief Summary

> **Abstract A.1.1**
>
> Multiplicative metric space was introduced in [Agamirza E Bashirov et.al, Multiplicative Calculus and its Applications, J. Math. Anal. Appl. 337 (2008) 36–48]. Combining this idea with Saks space, see references [1-4], [10-12] and [16] contained in [Keun Saeng Park, A Common Fixed Point Theorem in Saks Spaces, Journal of the Korea Society of Mathematical Education, Dec. 1982, Vol. XXI.No.1] we introduce multiplicative Saks space, and generalize multiplicative analogue of weak contractions by defining them implicitly via multiplicative c-class functions [Clement Ampadu and Arslan Hojat Ansari, FIXED POINT THEOREMS IN COMPLETE MULTIPLICATIVE METRIC SPACES WITH APPLICATION TO MULTIPLICATIVE ANALOGUE OF C-CLASS FUNCTIONS, JP Journal of Fixed Point Theory and Applications, August 2016, Volume 11, Issue 2, Pages 113 - 124]. Consequently, we obtain fixed point theorems for mappings satisfying such contractive conditions.

1.2 Preliminaries

> **Definition A.1.1**
>
> Let X be a linear space. A real valued function f^* defined on X will be called a multiplicative B-norm if it satisfies the following conditions
>
> (a) $f^*(x) = 1$ iff $x = 1$
>
> (b) $f^*(x+y) \leq f^*(x) \cdot f^*(y)$
>
> (c) $f^*(ax) = f^*(x)^{|a|}$, where a is any real number

> **Definition A.2 1**
>
> Let X be a linear space. A real valued function f^* defined on X will be called a multiplicative F-norm if it satisfies (a) and (b) of previous definition and the following: If the sequence $\{a_n\}$ of real numbers multiplicative converges to a, then $\lim_{n\to\infty} f^*(a_n x_n - ax) = 1$

> **Remark A.3 1**
>
> (X, N_1^*, N_2^*) will denote a multiplicative two-norm space, where X is a linear space, N_1^* is a multiplicative B-norm, and N_2^* is a multiplicative F-norm

> **Definition A.4 1**
>
> Let N_1^* and N_2^* be defined on X, and suppose $\{x_n\}$ is a sequence in X such that $\lim_{n\to\infty} N_1^*(x_n) = 1$ implies $\lim_{n\to\infty} N_2^*(x_n) = 1$, then we say N_1^* is non-weaker than N_2^* and write $N_2^* \leq N_1^*$. Furthermore, we say N_1^* and N_2^* are equivalent if $N_2^* \leq N_1^*$ and $N_1^* \leq N_2^*$

> **Definition A.5 1**
>
> Let (X, N_1^*, N_2^*) be a multiplicative two-norm space. A sequence $\{x_n\}$ in X will be called multiplicative convergent to $x \in X$ if $\sup_n N_1^*(x_n) < \infty$ and
> $$\lim_{n\to\infty} N_2^*(x_n - x) = 1$$

> **Definition A.6 1**
>
> Let (X, N_1^*, N_2^*) be a multiplicative two-norm space. A sequence $\{x_n\}$ in X will be called multiplicative Cauchy if $\lim_{n,m\to\infty} N_2^*(x_n - x_m) = 1$

> **Definition A.7 1**
>
> Let (X, N_1^*, N_2^*) be a multiplicative two-norm space. We will say this space is multiplicative complete if every multiplicative Cauchy sequence in (X, N_1^*, N_2^*) is a multiplicative convergent sequence in (X, N_1^*, N_2^*)

> **Definition A.8 1**
>
> Let N_1^* and N_2^* be defined on a linear space X. Put $X_s^* = \{x \in X : N_1^*(x) < a \text{ for some } a > 1\}$ and $m(x,y) = N_2^*(x-y)$ for all $x, y \in X_s^*$, then m is a multiplicative metric on X_s^* and the multiplicative metric space (X_s^*, m) will be called a multiplicative Saks set

> **Definition A.9 1**
>
> A complete multiplicative Saks set (X_s^*, m) will be called a multiplicative Saks space and will be denoted by (X, N_1^*, N_2^*)

In the sequel we will need the following, which is multiplicative version of a lemma due to Orlicz [W. Orlicz: Linear operators in Saks spaces (I). Stud. Math. 11 (1950), 237–272]

Lemma A.10 1

Let $(X_s^*, m) = (X, N_1^*, N_2^*)$ be a multiplicative Saks space. Then the following are equivalent

(a) N_1^* is equivalent to N_2^* on X

(b) (X, N_1^*) is multiplicative version of Banach space and $N_1^* \leq N_2^*$ on X

(c) (X, N_2^*) is multiplicative version of Frechet space and $N_2^* \leq N_1^*$ on X

Definition A.11 1

A mapping $T : X \mapsto X$, where $(X_s^*, m) = (X, N_1^*, N_2^*)$ is a multiplicative Saks space will be called weakly-$(\phi - F)$-contractive if

$$N^*(Tx - Ty) \leq F\left(\frac{N_2^*(x-y)}{\phi(N_2^*(x-y))}\right)$$

where $x, y \in X$, $\phi : [1, \infty) \mapsto [1, \infty)$ is a continuous and nondecreasing function such that $\phi(t) = 1$ iff $t = 1$, and $F(\frac{x}{y}) := F(x, y)$ is a multiplicative C-class function [Clement Ampadu and Arslan Hojat Ansari, FIXED POINT THEOREMS IN COMPLETE MULTIPLICATIVE METRIC SPACES WITH APPLICATION TO MULTIPLICATIVE ANALOGUE OF C-CLASS FUNCTIONS, JP Journal of Fixed Point Theory and Applications, August 2016, Volume 11, Issue 2, Pages 113 - 124]

Taking inspiration from [Khan M.S., Swaleh M., Sessa S., Fixed points theorems by altering distances between the points, Bull. Austral. Math. Soc., 30(1984), 1-9] we introduce the following

Definition A.12 1

A function $\psi : [1, \infty) \mapsto [1, \infty)$ will be called a modified altering distance function if the following properties are satisfied

(a) ψ is monotone increasing and continuous

(b) $\psi(t) = 1$ iff $t = 1$

1.3 Main Results

Theorem A.1 1

Let $(X_s^*, m) = (X, N_1^*, N_2^*)$ be a multiplicative Saks space in which N_1^* is equivalent to N_2^* on X. Let $T : X \mapsto X$ be a self-mapping which satisfies the following inequality

$$\Psi(N_2^*(Tx - Ty)) \leq F\left(\frac{\Psi(M^*(x,y))}{\Phi(N^*(x,y))}\right)$$

where $x, y \in X$, $x \neq y$

$$M^*(x,y) = \max\{N_2^*(x-y), (N_2^*(x-Tx) \cdot N_2^*(y-Ty))^{\frac{1}{2}}, (N_2^*(y-Tx) \cdot N_2^*(x-Ty))^{\frac{1}{2}}\}$$

$$N^*(x,y) = \min\{N_2^*(x-y), (N_2^*(y-Tx) \cdot N_2^*(x-Ty))^{\frac{1}{2}}\}$$

$\Phi : [1, \infty) \mapsto [1, \infty)$ is a lower semi continuous function with $\Phi(t) > 1$ for all $t > 1$, $\Phi(1) = 1$ and $\Psi : [1, \infty) \mapsto [1, \infty)$ is a modified altering distance function which in addition is strictly monotone increasing, $F(x,y) := F(\frac{x}{y})$ is a multiplicative C-class function [Clement Ampadu and Arslan Hojat Ansari, FIXED POINT THEOREMS IN COMPLETE MULTIPLICATIVE METRIC SPACES WITH APPLICATION TO MULTIPLICATIVE ANALOGUE OF C-CLASS FUNCTIONS, JP Journal of Fixed Point Theory and Applications, August 2016, Volume 11, Issue 2, Pages 113- 124]. Then there is a unique fixed point of T

> **Proof of Theorem A.1 1**
>
> Let $x_0 \in X$. Define a sequence $\{x_n\}$ in X by $x_{n+1} = Tx_n$ for all $n \in \mathbb{N} \cup \{0\}$. If $x_n = x_{n+1}$, then x_n is a fixed point of T. Hence for all $n \in \mathbb{N} \cup \{0\}$, we assume that $x_n \neq x_{n+1}$. From the contractive definition of the theorem we have,
>
> $$\Psi(N_2^*(x_{n+1} - x_{n+2})) = \Psi(N_2^*(Tx_n - Tx_{n+1})) \\ \leq F\left(\frac{\Psi(M^*(x_n, x_{n+1}))}{\Phi(N^*(x_n, x_{n+1}))}\right) \quad (1.1)$$
>
> where
>
> $$M^*(x_n, x_{n+1}) = \max\left\{N_2^*(x_n - x_{n+1}), \left(N_2^*(x_n - x_{n+1}) \cdot N_2^*(x_{n+1} - x_{n+2})\right)^{\frac{1}{2}},\right. \\ \left. (N_2^*(x_{n+1} - x_{n+1}) \cdot N_2^*(x_n - x_{n+2}))^{\frac{1}{2}}\right\} \quad (1.2)$$
>
> and
>
> $$N^*(x_n, x_{n+1}) = \min\left\{N_2^*(x_n - x_{n+1}), \left(N_2^*(x_{n+1} - x_{n+1}) \cdot N_2^*(x_n - x_{n+2})\right)^{\frac{1}{2}}\right\} \quad (1.3)$$
>
> If possible, let for some n, $N_2^*(x_n - x_{n+1}) < N_2^*(x_{n+1} - x_{n+2})$. By multiplicative triangle inequality we have
>
> $$1 < \frac{N_2^*(x_{n+1} - x_{n+2})}{N_2^*(x_n - x_{n+1})} \leq N_2^*(x_n - x_{n+2})$$
>
> Since $x_n \neq x_{n+1}$, it follows that $N_2^*(x_n - x_{n+1}) > 1$. From (1.1), (1.2), and (1.3) and our assumption, we have
>
> $$\Psi(N_2^*(x_{n+1} - x_{n+2})) \leq F\left(\frac{\Psi(N_2^*(x_{n+1} - x_{n+2}))}{\Phi(N_2^*(x_n, x_{n+1}))}\right) \\ < \Psi(N_2^*(x_{n+1} - x_{n+2}))$$
>
> which is a contradiction. Hence for all $n \in \mathbb{N} \cup \{0\}$, we have,
>
> $$N_2^*(x_{n+1} - x_{n+2}) \leq N_2^*(x_n - x_{n+1})$$
>
> From the above, we obtain from (1.2) and (1.3), for all $n \in \mathbb{N} \cup \{0\}$,
>
> $$M^*(x_n, x_{n+1}) = N_2^*(x_n - x_{n+1}) \quad (1.4)$$
>
> $$N^*(x_n, x_{n+1}) = N_2^*(x_n - x_{n+2})^{\frac{1}{2}} \quad (1.5)$$
>
> Using (1.4) and (1.5) in (1.1), we have,
>
> $$\Psi(N_2^*(x_{n+1} - x_{n+2})) \leq F\left(\frac{\Psi(N_2^*(x_n - x_{n+1}))}{\Phi(N_2^*(x_n - x_{n+2})^{\frac{1}{2}})}\right) \quad (1.6)$$
>
> Since, $N_2^*(x_{n+1} - x_{n+2}) \leq N_2^*(x_n - x_{n+1})$, it follows that the sequence $\{N_2^*(x_n - x_{n+1})\}$ is a monotone decreasing sequence of non-negative real numbers. Hence there exists $r \geq 1$ such that $\lim_{n \to \infty} N_2^*(x_n - x_{n+1}) = r$. Now observe that
>
> $$\frac{N_2^*(x_n - x_{n+2})}{r^2} \leq \frac{N_2^*(x_n, x_{n+1}) \cdot N_2^*(x_{n+1} - x_{n+2})}{r^2}$$

> **Proof of Theorem A.1 continued 1**
>
> It follows that
>
> $$\frac{(\max\{N_2^*(x_n - x_{n+2}), 2r\})^2}{N_2^*(x_n - x_{n+2}) + 2r} \leq \frac{(\max\{N_2^*(x_n - x_{n+1}), r\})^2}{N_2^*(x_n - x_{n+1}) + r} \cdot \frac{(\max\{N_2^*(x_{n+1} - x_{n+2}), r\})^2}{N_2^*(x_{n+1} - x_{n+2}) + r} \to 1$$
>
> as $n \to \infty$. Consequently, $\lim_{n \to \infty} N_2^*(x_n - x_{n+2}) = r^2$. Taking limits in (1.6) and using the properties of Ψ, Φ and F, we have, $\Psi(r) \leq F\left(\frac{\Psi(r)}{\Phi(r^2)}\right)$, which implies $r = 1$. Hence we have,
>
> $$\lim_{n \to \infty} N_2^*(x_n - x_{n+1}) = 1 \tag{1.7}$$
>
> Next we show that $\{x_n\}$ is a multiplicative Cauchy sequence. If otherwise there exist $\epsilon > 1$ and sequences of natural numbers $\{m(k)\}$ and $\{n(k)\}$ such that for every natural number k
>
> $$n(k) > m(k) > k \tag{1.8}$$
>
> and
>
> $$N_2^*(x_{m(k)} - x_{n(k)}) \geq \epsilon \tag{1.9}$$
>
> Corresponding to $m(k)$ we can choose $n(k)$ to be the smallest integer such that the inequality immediately above holds. Thus, we have,
>
> $$N_2^*(x_{m(k)} - x_{n(k)-1}) < \epsilon \tag{1.10}$$
>
> Note that (1.9) implies that $N_2^*(Tx_{m(k)-1} - Tx_{n(k)-1}) \neq 1$. Hence $x_{m(k)-1} \neq x_{n(k)-1}$. Putting $x = x_{m(k)-1}$ and $y = x_{n(k)-1}$ in the contractive definition of the theorem, we have,
>
> $$\Psi(N_2^*(x_{m(k)} - x_{n(k)})) = \Psi(N_2^*(Tx_{m(k)-1} - Tx_{n(k)-1}))$$
> $$\leq F\left(\frac{\Psi(M^*(x_{m(k)-1}, x_{n(k)-1}))}{\Phi(N^*(x_{m(k)-1}, x_{n(k)-1}))}\right) \tag{1.11}$$
>
> where
>
> $$M^*(x_{m(k)-1}, x_{n(k)-1}) = \max\{N_2^*(x_{m(k)-1} - x_{n(k)-1}),$$
> $$(N_2^*(x_{m(k)-1} - x_{m(k)}) \cdot N_2^*(x_{n(k)-1} - x_{n(k)}))^{\frac{1}{2}}, \tag{1.12}$$
> $$(N_2^*(x_{n(k)-1} - x_{m(k)}) \cdot N_2^*(x_{m(k)-1} - x_{n(k)}))^{\frac{1}{2}}\}$$
>
> and
>
> $$N^*(x_{m(k)-1}, x_{n(k)-1}) = \min\{N_2^*(x_{m(k)-1} - x_{n(k)-1}),$$
> $$(N_2^*(x_{n(k)-1} - x_{m(k)}) \cdot N_2^*(x_{m(k)-1} - x_{n(k)}))^{\frac{1}{2}}\} \tag{1.13}$$
>
> Thus for every positive integer k we have
>
> $$\epsilon \leq N_2^*(x_{m(k)} - x_{n(k)})$$
> $$\leq N_2^*(x_{m(k)} - x_{n(k)-1}) \cdot N_2^*(x_{n(k)-1} - x_{n(k)})$$
> $$< \epsilon \cdot N_2^*(x_{n(k)-1} - x_{n(k)})$$
>
> Taking limits, we get
>
> $$\lim_{k \to \infty} N_2^*(x_{m(k)} - x_{n(k)}) = \epsilon \tag{1.14}$$
>
> Now observe that
>
> $$N_2^*(x_{m(k)-1} - x_{n(k)-1}) \leq N_2^*(x_{m(k)-1} - x_{m(k)})$$
> $$\cdot N_2^*(x_{m(k)} - x_{n(k)}) \cdot N_2^*(x_{n(k)} - x_{n(k)-1})$$

> **Proof of Theorem A.1 continued 1**
>
> and
> $$N_2^*(x_{m(k)} - x_{n(k)}) \leq N_2^*(x_{m(k)} - x_{m(k)-1}) \\ \cdot N_2^*(x_{m(k)-1} - x_{n(k)-1}) \cdot N_2^*(x_{n(k)-1} - x_{n(k)})$$
>
> Taking limits in the above two inequalities we deduce that
>
> $$\lim_{k \to \infty} N_2^*(x_{n(k)-1} - x_{m(k)}) = \epsilon \qquad (1.15)$$
>
> Taking limits in (1.11) and using properties of Ψ, Φ and F, we deduce that
>
> $$\Psi(\epsilon) \leq F\left(\frac{\Psi(\epsilon)}{\Phi(\epsilon)}\right)$$
>
> which is a contradiction since $\epsilon > 1$. Hence $\{x_n\}$ is a multiplicative Cauchy sequence with respect to N_1^*. From Lemma A.10, (X, N_1^*) is multiplicative version of Banach space, therefore there is $z \in X$ such that $\lim_{n \to \infty} x_n = z$. Since $x_n \neq x_{n+1}$ there is a subsequence $\{x_{n(k)}\}$ of $\{x_n\}$ such that $z \neq x_{n(k)}$ for all k. Now observe that
>
> $$\Psi(N_2^*(x_{n(k)+1} - Tz)) \leq F\left(\frac{\Psi(M^*(x_{n(k)}, z))}{\Phi(N^*(x_{n(k)}, z))}\right) \qquad (1.16)$$
>
> where
>
> $$M^*(x_{n(k)}, z) = \max\{N_2^*(x_{n(k)} - z), (N_2^*(x_{n(k)} - x_{n(k)+1}) \cdot N_2^*(z - Tz))^{\frac{1}{2}}, \\ (N_2^*(x_{n(k)} - Tz) \cdot N_2^*(z - x_{n(k)+1}))^{\frac{1}{2}}\} \qquad (1.17)$$
>
> and
>
> $$N^*(x_{n(k)}, z) = \min\{N_2^*(x_{n(k)} - z), (N_2^*(x_{n(k)} - Tz) \cdot N_2^*(z - x_{n(k)+1}))^{\frac{1}{2}}\} \qquad (1.18)$$
>
> Taking limits in the three inequalities immediately above, and using the properties of Φ, Ψ and F we deduce that $\Psi(N_2^*(z - Tz)) \leq \Psi((N_2^*(z - Tz))^{\frac{1}{2}})$, which is a contradiction unless $N_2^*(z - Tz) = 1$, that is, $z = Tz$. Hence z is a fixed point of T. If $z_1 \neq z_2$ are two fixed points of T, then we obtain
>
> $$\Psi(N_2^*(z_1 - z_2)) \leq F\left(\frac{\Psi(M^*(z_1, z_2))}{\Phi(N^*(z_1, z_2))}\right)$$
>
> which by the properties of Φ, Ψ and F implies $N_2^*(z_1 - z_2) = 1$, that is, $z_1 = z_2$ and uniqueness follows

1.4 Exercises

Exercise A.1 1

Taking inspiration from [B. E. Rhoades, Some theorems on weakly contractive maps, Nonlinear Analysis 47 (2001) 2683–2693] prove the following: Let $(X_s^*, m) = (X, N_1^*, N_2^*)$ be a multiplicative Saks space in which N_1^* is equivalent to N_2^* on X. Let $T : X \mapsto X$ be a self-mapping which satisfies the following inequality

$$N_2^*(Tx - Ty) \leq F\left(\frac{N_2^*(x-y)}{\Phi(N_2^*(x-y))}\right)$$

where $x, y \in X$, $x \neq y$, $\Phi : [1, \infty) \mapsto [1, \infty)$ is continuous, nondecreasing such that Φ is positive on $(1, \infty)$, $\Phi(1) = 1$, and $\lim_{t \to \infty} \Phi(t) = \infty$, $F(x, y) := F(\frac{x}{y})$ is a multiplicative C-class function [Clement Ampadu and Arslan Hojat Ansari, FIXED POINT THEOREMS IN COMPLETE MULTIPLICATIVE METRIC SPACES WITH APPLICATION TO MULTIPLICATIVE ANALOGUE OF C-CLASS FUNCTIONS, JP Journal of Fixed Point Theory and Applications, August 2016, Volume 11, Issue 2, Pages 113- 124]. Then T has a unique fixed point ζ in X

Exercise A.2 1

Let Ψ denote the class of all functions $\psi : [1, \infty) \mapsto [1, \infty)$ that are continuous nondecreasing, and such that $\psi^{-1}(\{1\}) = 1$. Also let Φ denote the class of all functions $\phi : [1, \infty) \mapsto [1, \infty)$ that are lower semi-continuous and such that $\phi^{-1}(\{1\}) = 1$. Taking inspiration from [I. Beg and M. Abbas, Coincidence point and invariant approximation for mappings satisfying generalized weak contractive condition, Fixed Point Theory Appl. (2006), Article ID 74503, 7 pp] prove the following:

Let $(X_s^*, m) = (X, N_1^*, N_2^*)$ be a multiplicative Saks space in which N_1^* is equivalent to N_2^* on X. Let $k, g : X \mapsto X$ be self-mappings which satisfies the following inequality

$$\psi(N_2^*(kx - ky)) \leq F\left(\frac{\psi(N_2^*(gx - gy))}{\phi(N_2^*(gx - gy))}\right)$$

where $x, y \in X$, $x \neq y$, $\psi \in \Psi$, $\phi \in \Phi$, $F(x, y) := F(\frac{x}{y})$ is a multiplicative C-class function [Clement Ampadu and Arslan Hojat Ansari, FIXED POINT THEOREMS IN COMPLETE MULTIPLICATIVE METRIC SPACES WITH APPLICATION TO MULTIPLICATIVE ANALOGUE OF C-CLASS FUNCTIONS, JP Journal of Fixed Point Theory and Applications, August 2016, Volume 11, Issue 2, Pages 113- 124]. If $kX \subset gX$, and gX is a complete subspace of X, then k, g have a coincidence point in X

Exercise A.3 1

Taking inspiration from [J. Harjani, B. López, K. Sadarangani, Fixed point theorems for weakly C-contractive mappings in ordered metric spaces, Computers and Mathematics with Applications 61 (2011) 790–796] prove the following:

Let $(X_s^*, m) = (X, N_1^*, N_2^*)$ be a multiplicative Saks space in which N_1^* is equivalent to N_2^* on X, and suppose $(X, \leq)$ is a partially ordered set. Let $T : X \mapsto X$ be a continuous and nondecreasing mapping such that for $x \geq y$

$$\psi(N_2^*(Tx - Ty)) \leq F\left(\frac{\sqrt{N_2^*(x - Ty) \cdot N_2^*(y - Tx)}}{\phi(N_2^*(x - Ty), N_2^*(y - Ty))}\right)$$

where $x, y \in X$, $x \neq y$, $\phi : [1, \infty)^2 \mapsto [1, \infty)$ is a continuous function such that $\phi(x, y) = 1$ iff $x = y = 1$, $F(x, y) := F(\frac{x}{y})$ is a multiplicative C-class function [Clement Ampadu and Arslan Hojat Ansari, FIXED POINT THEOREMS IN COMPLETE MULTIPLICATIVE METRIC SPACES WITH APPLICATION TO MULTIPLICATIVE ANALOGUE OF C-CLASS FUNCTIONS, JP Journal of Fixed Point Theory and Applications, August 2016, Volume 11, Issue 2, Pages 113- 124]. If there exists $x_0 \in X$ with $x_0 \leq Tx_0$, then T has a fixed point

Exercise A.4 1

Taking inspiration from [Q. Zhang, Y. Song, Fixed point theory for ϕ-weak contractions, Appl. Math. Lett., 22(1), (2009), 75-78] prove the following:

Let $(X_s^*, m) = (X, N_1^*, N_2^*)$ be a multiplicative Saks space in which N_1^* is equivalent to N_2^* on X. Let $T, S : X \mapsto X$ be two mappings such that for all $x, y \in X$ we have

$$N_2^*(Tx - Sy) \leq F\left(\frac{M(x, y)}{\phi(M(x, y))}\right)$$

where $\phi : [1, \infty) \mapsto [1, \infty)$ is a lower semi-continuous function and $\phi(t) = 1$ iff $t = 1$, $F(x, y) := F(\frac{x}{y})$ is a multiplicative C-class function [Clement Ampadu and Arslan Hojat Ansari, FIXED POINT THEOREMS IN COMPLETE MULTIPLICATIVE METRIC SPACES WITH APPLICATION TO MULTIPLICATIVE ANALOGUE OF C-CLASS FUNCTIONS, JP Journal of Fixed Point Theory and Applications, August 2016, Volume 11, Issue 2, Pages 113- 124], and

$$M(x, y) = \max\{N_2^*(x - y), N_2^*(Tx - x), N_2^*(Sy - y), \sqrt{N_2^*(y - Tx) \cdot N_2^*(x - Sy)}\}$$

Then there exists a unique point $u \in X$ such that $u = Tu = Su$

1.5 References

(1) Agamirza E Bashirov et.al, Multiplicative Calculus and its Applications, J. Math. Anal. Appl. 337 (2008) 36–48

(2) Keun Saeng Park, A Common Fixed Point Theorem in Saks Spaces, Journal of the Korea Society of Mathematical Education, Dec. 1982, Vol. XXI.No.1

(3) Clement Ampadu and Arslan Hojat Ansari, FIXED POINT THEOREMS IN COMPLETE MULTIPLICATIVE METRIC SPACES WITH APPLICATION TO MULTIPLICATIVE ANALOGUE OF C-CLASS FUNCTIONS, JP Journal of Fixed Point Theory and Applications, August 2016, Volume 11, Issue 2, Pages 113 - 124

(4) W. Orlicz: Linear operators in Saks spaces (I). Stud. Math. 11 (1950), 237–272

(5) Khan M.S., Swaleh M., Sessa S., Fixed points theorems by altering distances between the points, Bull. Austral. Math. Soc., 30(1984), 1-9

(6) B. E. Rhoades, Some theorems on weakly contractive maps, Nonlinear Analysis 47 (2001) 2683–2693

(7) I. Beg and M. Abbas, Coincidence point and invariant approximation for mappings satisfying generalized weak contractive condition, Fixed Point Theory Appl. (2006), Article ID 74503, 7 pp

(8) J. Harjani, B. López, K. Sadarangani, Fixed point theorems for weakly C-contractive mappings in ordered metric spaces, Computers and Mathematics with Applications 61 (2011) 790–796

(9) Q. Zhang, Y. Song, Fixed point theory for ϕ-weak contractions, Appl. Math. Lett., 22(1), (2009), 75-78

Chapter 2

Coincidence and Common Fixed Point Theorems for Implicit f-Weak Multiplicative Contractions in Multiplicative Cone Metric Space

2.1 Brief Summary

Abstract B.1 1

The f-weak contractions are generalization of weak contractions [Ya. I. Alber and S. Guerre-Delabriere, Principles of weakly contractive maps in Hilbert spaces, in: I. Gohberg and Yu Lyubich *New Results in Operator Theory and it's Applications*, in: Oper. Theory Adv. Appl. Birkhauser, Basel, 1997]. In this chapter we introduce implicit f-weak multiplicative contractions and establish coincidence and common fixed point results for such contractions in multiplicative cone metric space.

2.2 Preliminaries

Definition B.1 1

[Clement Boateng Ampadu. A COUPLED VERSION OF THE HIGHER-ORDER BANACH CONTRACTION PRINCIPLE IN MULTIPLICATIVE CONE METRIC SPACE. JP Journal of Applied Mathematics, To Appear] Let E be a real Banach space and P be a subset of E. P is said to be a multiplicative cone if

(a) P is closed, nonempty, and satisfies $P \neq \{1\}$

(b) $x^a \cdot x^b \in P$ for all $x, y \in P$ and non-negative real numbers a, b

(c) $x \in P$ and $\frac{1}{x} \in P$ imply $x = 1$, that is, $P \cap \frac{1}{P} = 1$

Definition B.2 1

[Clement Boateng Ampadu. A COUPLED VERSION OF THE HIGHER-ORDER BANACH CONTRACTION PRINCIPLE IN MULTIPLICATIVE CONE METRIC SPACE. JP Journal of Applied Mathematics, To Appear] For a given multiplicative cone $P \subseteq E$, the partial ordering $\leq$ with respect to P is defined by $x \leq y$ iff $\frac{y}{x} \in P$. $x \ll y$ will stand for $\frac{y}{x} \in int(P)$, where $int(P)$ denotes the interior of P. $x < y$ will indicate $x \leq y$ but $x \neq y$

Definition B.3 1

[Clement Boateng Ampadu. A COUPLED VERSION OF THE HIGHER-ORDER BANACH CONTRACTION PRINCIPLE IN MULTIPLICATIVE CONE METRIC SPACE. JP Journal of Applied Mathematics, To Appear] The multiplicative cone P will be called multiplicative normal if there exists a constant $M > 0$ such that for every $x, y \in E$ if $1 \leq x \leq y$, then, $\|x\| \leq \|y\|^M$. The least positive number satisfying this inequality will be called the multiplicative normal constant of P

Definition B.4 1

The multiplicative cone P will be called regular iff every decreasing sequence bounded from below is multiplicative convergent

Remark B.5 1

In the sequel we will always assume E is a Banach space, P is a multiplicative cone in E with $intP \neq \emptyset$ and $\leq$ is a partial ordering with respect to P

Definition B.6 1

[Clement Boateng Ampadu. A COUPLED VERSION OF THE HIGHER-ORDER BANACH CONTRACTION PRINCIPLE IN MULTIPLICATIVE CONE METRIC SPACE. JP Journal of Applied Mathematics, To Appear] Let X be a nonempty set and let E be a real Banach space equipped with the partial ordering $\leq$ with respect to the multiplicative cone $P \subset E$. Suppose that the mapping $m : X \times X \mapsto E$ satisfies the following conditions

(a) $1 \leq m(x, y)$ for all $x, y \in X$ and $m(x, y) = 1$ iff $x = y$

(b) $m(x, y) = m(y, x)$ for all $x, y \in X$

(c) $m(x, y) \leq m(x, z) \cdot m(z, y)$ for all $x, y, z \in X$

Then m is called a multiplicative cone metric on X, and (X, m) is called a multiplicative cone metric space

Remark B.7 1

[Clement Boateng Ampadu. A COUPLED VERSION OF THE HIGHER-ORDER BANACH CONTRACTION PRINCIPLE IN MULTIPLICATIVE CONE METRIC SPACE. JP Journal of Applied Mathematics, To Appear] Convergence of a sequence, Cauchy sequence and Completeness in multiplicative cone metric space are defined the same way as in generalized multiplicative cone b-metric space (see Definition A.11 and Definition A.12 of Chapter 15 [Ampadu, Clement (2016). Fixed Point Theory for Higher-Order Mappings. lulu.com. ISBN:5800118959925])

The following is the multiplicative version of a Lemma contained in [L.-G. Huang and X. Zhang, Cone metric spaces and fixed point theorems of contractive mappings, J.Math.Anal.Appl. 332 (2007) 1468-1476]

> **Lemma B.8 1**
>
> Let (X, m) be a multiplicative cone metric space, P be a multiplicative normal cone with multiplicative normal constant K. Let $\{x_n\}$ be a sequence in X. Then $\{x_n\}$ multiplicative converges to x iff $m(x_n, x) \to 1$, as $n \to \infty$

The following is the multiplicative version of a Lemma contained in [L.-G. Huang and X. Zhang, Cone metric spaces and fixed point theorems of contractive mappings, J.Math.Anal.Appl. 332 (2007) 1468-1476] which shows that the limit of sequences in multiplicative cone metric space is unique

> **Lemma B.9 1**
>
> Let (X, m) be a multiplicative cone metric space, P be a multiplicative normal cone with multiplicative normal constant K. Let $\{x_n\}$ be a sequence in X. If $\{x_n\}$ multiplicative converges to x and $\{x_n\}$ multiplicative converges to y, then $x = y$

The following is the multiplicative version of a Lemma contained in [L.-G. Huang and X. Zhang, Cone metric spaces and fixed point theorems of contractive mappings, J.Math.Anal.Appl. 332 (2007) 1468-1476]

> **Lemma B.10 1**
>
> Let (X, m) be a multiplicative cone metric space and $\{x_n\}$ be a sequence in X. If $\{x_n\}$ multiplicative converges to $x \in X$, then $\{x_n\}$ is a multiplicative Cauchy sequence

The following is the multiplicative version of a Lemma contained in [L.-G. Huang and X. Zhang, Cone metric spaces and fixed point theorems of contractive mappings, J.Math.Anal.Appl. 332 (2007) 1468-1476]

> **Lemma B.11 1**
>
> Let (X, m) be a multiplicative cone metric space, P be a multiplicative normal cone with multiplicative normal constant K. Let $\{x_n\}$ be a sequence in X. Then $\{x_n\}$ is a multiplicative Cauchy sequence iff $m(x_n, x_k) \to 1$ as $n, k \to \infty$

The following is the multiplicative version of a Lemma contained in [L.-G. Huang and X. Zhang, Cone metric spaces and fixed point theorems of contractive mappings, J.Math.Anal.Appl. 332 (2007) 1468-1476]

> **Lemma B.12 1**
>
> Let (X, m) be a multiplicative cone metric space, P be a multiplicative normal cone with multiplicative normal constant K. Let $\{x_n\}$ and $\{y_n\}$ be two sequences in X such that $\lim_{n \to \infty} x_n = x$ and $\lim_{n \to \infty} y_n = y$. Then $\lim_{n \to \infty} m(x_n, y_n) = m(x, y)$

The following is the multiplicative version of a Lemma contained in [D. Ilic, V. Rakocevic, Quasi-contraction on a cone metric space, Appl. Math. Lett. 22 (2009) 728-731]

> **Lemma B.13 1**
>
> Let P be a multiplicative normal cone in E, then
>
> (a) if $1 \leq x \leq y$ and $a \geq 1$, where a is a real number, then $1 \leq x^a \leq y^a$
>
> (b) if $1 \leq x_n \leq y_n$, for $n \in \mathbb{N}$, and $x_n \to x$, $y_n \to y$, then $1 \leq x \leq y$

From [G. Jungck, S. Radenovic, S. Radojevic, V. Rakocevic, Common fixed point theorems for weakly compatible pairs on cone metric spaces, Fixed Point Theory Appl. (2009). Art. ID 643849, 13 pp.] we deduce the following

Lemma B.14 1

Let E be a real Banach space with multiplicative cone P in E, then for $a, b, c \in E$

(a) if $a \leq b$ and $b \ll c$, then $a \ll c$

(b) if $a \ll b$ and $b \ll c$, then $a \ll c$

Definition B.15 1

[J. Jachymski, Order-Theoretic Aspects of Metric Fixed Point Theory, Handbook of Metric Fixed Point Theory, Kluwer Academic Publishers, Dordrecht, 2001] Let $(Y, \leq)$ be a partially ordered set. Then a function $F : Y \mapsto Y$ is said to be monotone increasing if it preserves ordering, that is, given $x, y \in Y$, $x \leq y$ implies $Fx \leq Fy$

Definition B.16 1

[W. SINTUNAVARAT AND P. KUMAM, COMMON FIXED POINTS OF f-WEAK CONTRACTIONS IN CONE METRIC SPACES, Bulletin of the Iranian Mathematical Society Vol. 38 No. 2 (2012), pp 293-303] Let f and T be self mappings of a nonempty set X. If $w = fx = Tx$ for some $x \in X$, then x is called a coincidence point of f and T, and w is called a point of coincidence of f and T. If $w = x$, then x is called a common fixed point of f and T.

Using the multiplicative c-class function [Clement Ampadu and Arslan Hojat Ansari, FIXED POINT THEOREMS IN COMPLETE MULTIPLICATIVE METRIC SPACES WITH APPLICATION TO MULTIPLICATIVE ANALOGUE OF C-CLASS FUNCTIONS, JP Journal of Fixed Point Theory and Applications, August 2016, Volume 11, Issue 2, Pages 113 - 124] we define the f-weak contraction implicitly in the setting of multiplicative metric space as follows.

Definition B.17 1

Let (X, m) be a multiplicative metric space and $f : X \mapsto X$. A map $T : X \mapsto X$ will be called an $(\phi - F - f)$-weak multiplicative contraction if

$$m(Tx, Ty) \leq F\left[\frac{m(fx, fy)}{\phi(m(fx, fy))}\right]$$

for $x, y \in X$, where $\phi : [1, \infty) \mapsto [1, \infty)$ is a continuous and nondecreasing function with $\phi(t) = 1$ iff $t = 1$, and $F(x, y) := F(\frac{x}{y})$ is a multiplicative c-class function [Clement Ampadu and Arslan Hojat Ansari, FIXED POINT THEOREMS IN COMPLETE MULTIPLICATIVE METRIC SPACES WITH APPLICATION TO MULTIPLICATIVE ANALOGUE OF C-CLASS FUNCTIONS, JP Journal of Fixed Point Theory and Applications, August 2016, Volume 11, Issue 2, Pages 113 - 124]

2.3 Main Results

Theorem B.1 1

Let (X, m) be a multiplicative cone metric space with regular multiplicative cone P such that $m(x,y) \in int(P)$, for $x, y \in X$ with $x \neq y$. Let $f : X \mapsto X$ and $T : X \mapsto X$ be a mapping satisfying the inequality

$$m(Tx, Ty) \leq F\left[\frac{m(fx, fy)}{\psi(m(fx, fy))}\right]$$

for $x, y \in X$, where $\psi : int(P) \cup \{1\} \mapsto int(P) \cup \{1\}$ is a continuous and monotone increasing function with

(a) $\psi(t) = 1$ iff $t = 1$

(b) $\psi(t) \ll t$, for $t \in int(P)$

(c) either $\psi(t) \leq m(fx, fy)$ or $m(fx, fy) \leq \psi(t)$, for $t \in int(P) \cup \{1\}$ and $x, y \in X$

and $F(x, y) := F(\frac{x}{y})$ is a multiplicative c-class function [Clement Ampadu and Arslan Hojat Ansari, FIXED POINT THEOREMS IN COMPLETE MULTIPLICATIVE METRIC SPACES WITH APPLICATION TO MULTIPLICATIVE ANALOGUE OF C-CLASS FUNCTIONS, JP Journal of Fixed Point Theory and Applications, August 2016, Volume 11, Issue 2, Pages 113 - 124]. If $TX \subseteq fX$ and fX is a complete subspace of X, then f and T have a unique point of coincidence in X. Moreover, f and T have a unique common fixed point in X if $ffz = fz$ for some coincidence point z

Proof of Theorem B.1 1

Let $x_0 \in X$. Since $TX \subseteq fX$, we construct the sequence $\{fx_n\}$, where $fx_n = Tx_{n-1}$, $n \geq 1$. If $fx_{n+1} = fx_n$, for some n, then trivially f and T have a coincidence point in X. If $fx_{n+1} \neq fx_n$, for $n \in \mathbb{N}$, by the given condition, we have,

$$m(fx_n, fx_{n+1}) = m(Tx_{n-1}, Tx_n)$$
$$\leq F\left[\frac{m(fx_{n-1}, fx_n)}{\psi(fx_{n-1}, fx_n))}\right]$$

By the properties of ψ and F we deduce that $1 \leq \psi(t)$, for all $t \in int(P) \cup \{1\}$, thus

$$m(fx_n, fx_{n+1}) \leq m(fx_{n-1}, fx_n)$$

It follows that the sequence $\{m(fx_n, fx_{n+1})\}$ is monotonically decreasing. Since the multiplicative cone P is regular and $1 \leq m(fx_n, fx_{n+1})$, for all $n \in \mathbb{N}$, there exists $r \geq 1$ such that $m(fx_n, fx_{n+1}) \to r$ as $n \to \infty$. By continuity of ψ and F, and since

$$m(fx_n, fx_{n+1}) \leq F\left[\frac{m(fx_{n-1}, fx_n)}{\psi(fx_{n-1}, fx_n))}\right]$$

If we take limits in the above inequality we deduce that $r \leq F\left[\frac{r}{\psi(r)}\right]$ which is a contradiction unless $r = 1$. Therefore, $m(fx_n, fx_{n+1}) \to 1$ as $n \to \infty$. Let $c \in E$ with $1 \ll c$ be arbitrary. Since $m(fx_n, fx_{n+1}) \to 1$ as $n \to \infty$, there exists $k \in \mathbb{N}$ such that

$$m(fx_k, fx_{k+1}) \ll \psi(\psi(\sqrt{c}))$$

Let $B(fx_k, c) = \{fx \in X : m(fx_k, fx) \ll c\}$. Clearly $fx_k \in B(fx_k, c)$. Therefore, $B(fx_m, c)$ is nonempty. Now we will show that $Tx \in B(fx_k, c)$ for $fx \in B(fx_k, c)$. Let $x \in B(fx_k, c)$. By properties of ψ and F we consider two cases

<u>Case I:</u> $m(fx, fx_k) \leq \psi(\sqrt{c})$

In this case we deduce the following

$$m(Tx, fx_k) \leq m(Tx, Tx_k) \cdot m(Tx_k, fx_k)$$
$$\leq F\left[\frac{m(fx, fx_k)}{\psi(m(fx, fx_k))}\right] \cdot m(Tx_k, fx_k)$$
$$\leq F\left[\frac{m(fx, fx_k)}{\psi(m(fx, fx_k))}\right] \cdot m(fx_{k+1}, fx_k)$$
$$\leq m(fx, fx_k) \cdot m(fx_{k+1}, fx_k)$$
$$\leq \psi(\sqrt{c}) \cdot \psi(\psi(\sqrt{c}))$$
$$\ll \psi(\sqrt{c}) \cdot \psi(\sqrt{c})$$
$$\ll \sqrt{c} \cdot \sqrt{c}$$
$$= c$$

Proof of Theorem B.1 Continued 1

<u>Case II:</u> $\psi(\sqrt{c}) < m(fx, fx_k) \ll c$

$$m(Tx, fx_k) \leq m(Tx, Tx_k) \cdot m(Tx_k, fx_k)$$
$$\leq F\left[\frac{m(fx, fx_k)}{\psi(m(fx, fx_k))}\right] \cdot m(Tx_k, fx_k)$$
$$\leq F\left[\frac{m(fx, fx_k)}{\psi(m(fx, fx_k))}\right] \cdot m(fx_{k+1}, fx_k)$$
$$\leq \frac{m(fx, fx_k)}{\psi(\psi(\sqrt{c}))} \cdot \psi(\psi(\sqrt{c}))$$
$$\leq m(fx, fx_k)$$
$$\ll c$$

Therefore, T is a self-mapping of $B(fx_k, c)$. Since $fx_k \in B(fx_k, c)$ and $fx_n = Tx_{n-1}$, $n \geq 1$, it follows that $fx_n \in B(fx_k, c)$ for all $n \geq k$. Since c is arbitrary it follows that $\{fx_n\}$ is a multiplicative Cauchy sequence in fX. By multiplicative completeness of fX, there exists $x \in X$ such that $\lim_{n\to\infty} fx_n = fx$. Now observe that

$$m(fx_n, Tx) = m(Tx_{n-1}, Tx)$$
$$\leq F\left[\frac{m(fx_{n-1}, fx)}{\psi(m(fx_{n-1}, fx))}\right]$$

Taking limits in the above, we deduce that $m(fx, Tx) \leq 1$. Consequently, $m(fx, Tx) = 1$, that is, $fx = Tx$. Hence, x is a coincidence point of f and T. For the uniqueness of the coincidence point of f and T, let $y \in X$ ($y \neq x$) be another coincidence point of f and T. Observe that

$$m(fx, fy) = m(Tx, Ty)$$
$$\leq F\left[\frac{m(fx, fy)}{\psi(m(fx, fy))}\right]$$

From above, by the properties of ψ and F, we deduce $\psi(m(fx, fy)) \leq 1$, which is a contradiction. Therefore f and T have a unique point of coincidence in X. Now let z be a coincidence point of f and T. It follows from $ffz = fz$ and z being a coincidence point of f and T that $ffz = fz = Tz$. Now observe from the contractive condition of the theorem, and properties of ψ and F we have

$$m(Tfz, Tz) \leq F\left[\frac{m(ffz, fz)}{\psi(m(ffz, fz))}\right]$$
$$= F\left[\frac{1}{\psi(1)}\right]$$
$$= 1$$

Therefore $Tfz = Tz$, that is, $fz = ffz = Tfz$. Hence, fz is a common fixed point of f and T. Uniqueness of common fixed point follows from contractive condition of theorem.

Remark B.2 1

The multiplicative version of Theorem 2.1 [B. S. Choudhury and N. Metiya, Fixed points of weak contractions in cone metric spaces, Nonlinear Analysis 72 (2010), no. 3-4, 1589-1593] is obtained from the above theorem by taking f to be the identity.

Finally we have the following example which illustrates Theorem B.1

Example B.3 1

Let $X = [0,1]$, $E = \mathbb{R} \times \mathbb{R}$, with usual norm, be a real Banach space, $P = \{(x,y) \in E : x, y \geq 1\}$ be a regular multiplicative cone, and the partial ordering $\leq$ with respect to the multiplicative cone P be the usual partial ordering in E. Define $m : X \times X \mapsto E$ by $m(x,y) = (a^{|x-y|}, a^{|x-y|})$ for some $a > 1$ and $x, y \in X$. Then (X, m) is a complete multiplicative cone metric space with $m(x,y) \in int(P)$, for $x, y \in X$ with $x \neq y$. Let us define $\psi : int(P) \cup \{1\} \mapsto int(P) \cup \{1\}$ as

$$\psi(t) = (t_1^2, t_1^2), (t_1, t_2) \in int(P) \cup \{1\}, t_1 \leq t_2$$

$$\psi(t) = (t_2^2, t_2^2), (t_1, t_2) \in int(P) \cup \{1\}, t_1 > t_2$$

Clearly ψ has all the required properties. Let us define $f : X \mapsto X$ and $T : X \mapsto X$ as follows $fx = \sqrt{x}$ and $Tx = 1$. Without loss of generality, we may assume $x > y$ for $x, y \in X$, then we can check that

$$m(Tx, Ty) \leq F\left[\frac{m(fx, fy)}{\psi(m(fx, fy))}\right]$$

where $F(x,y) := F(\frac{x}{y}) = \frac{x}{y}$ is the multiplicative c-class function. Note that $1 \in X$ is the unique common fixed point of f and T

2.4 Exercises

Exercise B.1 1

Taking inspiration from [Binayak S. Choudhury, N. Metiya, Fixed points of weak contractions in cone metric spaces, Nonlinear Analysis 72 (2010) 1589-1593] give an example to show that if f is the identity in Theorem B.1, then it still holds

Exercise B.2 1

We will say $f, g : X \mapsto X$ form a weakly $F - (\phi, \psi)$ pair if $\phi(m(fx, fy)) \leq F\left[\frac{\phi(z)}{\psi(z)}\right]$ for some $z \in M_{f,g}(x,y)$, for all $x, y \in X$, where $M_{f,g}(x,y) = \{m(gx, gy), m(fx, gx), m(fy, gy)\}$, $\phi : P \mapsto P$ and $\psi : int(P) \cup \{1\} \mapsto int(P) \cup \{1\}$ are continuous functions with the following properties

(a) ϕ is strongly monotonic increasing

(b) $\phi(t) = 1 = \psi(t)$ iff $t = 1$

(c) $\psi(t) \ll t$, for $t \in int(P)$

(d) either $\psi(t) \leq m(x,y)$ or $m(x,y) \leq \psi(t)$, for $t \in int(P) \cup \{1\}$ and $x, y \in X$

and $F(x,y) := F(\frac{x}{y})$ is a multiplicative c-class function [Clement Ampadu and Arslan Hojat Ansari, FIXED POINT THEOREMS IN COMPLETE MULTIPLICATIVE METRIC SPACES WITH APPLICATION TO MULTIPLICATIVE ANALOGUE OF C-CLASS FUNCTIONS, JP Journal of Fixed Point Theory and Applications, August 2016, Volume 11, Issue 2, Pages 113 - 124].Taking inspiration from [C. T. AAGE AND J. N. SALUNKE, FIXED POINTS OF (ϕ, ψ)-WEAK CONTRACTIONS IN CONE METRIC SPACES, Ann. Funct. Anal. 2 (2011), no. 1, 59–71] prove the following: Let (X, m) be a multiplicative cone metric space with regular multiplicative cone P such that $m(x,y) \in int(P)$, for all $x, y \in X$ with $x \neq y$. Let $f, g : X \mapsto X$ form a weakly $F - (\phi, \psi)$ pair. If $f(X) \subset g(X)$ and $g(X)$ is a complete subspace of X, then f and g have a unique point of coincidence in X. Moreover, if f and g are weakly compatible, then f and g have a unique common fixed point in X

Exercise B.3 1

Taking inspiration from [BINAYAK S. CHOUDHURY, L. KUMAR, T. SOM; AND N. METIYA, A WEAK CONTRACTION PRINCIPLE IN PARTIALLY ORDERED CONE METRIC SPACE WITH THREE CONTROL FUNCTIONS, International Journal of Analysis and Applications, Volume 6, Number 1 (2014), 18-27] prove the following: Let $(X, \preceq)$ be a partially ordered set and suppose there exists a multiplicative cone metric m in X such that (X, m) is complete with regular multiplicative cone P such that $m(x,y) \in int(P)$, for $x, y \in X$ with $x \neq y$. Let $T : X \mapsto X$ be a continuous and nondecreasing mapping such that for all comparable $x, y \in X$

$$\psi(m(Tx, Ty)) \leq F\left[\frac{\gamma(m(x,y))}{\phi(m(x,y))}\right]$$

where $F(x,y) := F(\frac{x}{y})$ is a multiplicative c-class function [Clement Ampadu and Arslan Hojat Ansari, FIXED POINT THEOREMS IN COMPLETE MULTIPLICATIVE METRIC SPACES WITH APPLICATION TO MULTIPLICATIVE ANALOGUE OF C-CLASS FUNCTIONS, JP Journal of Fixed Point Theory and Applications, August 2016, Volume 11, Issue 2, Pages 113 - 124], and $\psi, \gamma, \phi : int(P) \cup \{1\} \mapsto int(P) \cup \{1\}$ are such that ψ, γ are continuous, ϕ is lower semi-continuous and also

(a) ψ is strongly monotonic increasing

(b) $\psi(t) = \gamma(t) = \phi(t) = 1$ iff $t = 1$

(c) $\frac{\psi(t)\phi(t)}{\gamma(t)} > 1$ for $t \in int(P)$

(d) $\phi(t) \ll t$, for $t \in int(P)$

(e) either $\phi(t) \leq m(x,y)$ or $m(x,y) \ll \phi(t)$, for $t \in int(P) \cup \{1\}$ and $x, y \in X$

If there exists $x_0 \in X$ such that $x_0 \preceq Tx_0$, then T has a fixed point in X

> **Exercise B.4 1**
>
> Let ψ and ϕ be defined as in Exercise B.2, and taking inspiration from [Hemant Kumar Nashine and Hassen Aydi, Common fixed points for generalized (ψ, ϕ)-weak contractions in ordered cone metric spaces, Applied General Topology, Volume 13, no. 2, 2012, pp. 151-166] prove the following: Let $(X, m, \preceq)$ be an ordered complete multiplicative cone metric space over a solid multiplicative cone P. Let $T, S, I, J : X \mapsto X$ be given mappings satisfying for every pair $(x, y) \in X \times X$ such that x and y are comparable
>
> $$\phi(m(Sx, Ty)) \leq F\left[\frac{\phi(\theta(x,y))}{\psi(\theta(x,y))}\right]$$
>
> where $\theta(x,y) \in \{m(Ix, Jy), \sqrt{(m(Ix, Sx) \cdot m(Jy, Ty)}, \sqrt{(m(Ix, Ty) \cdot m(Jy, Sx)}\}$, and $F(x,y) := F(\frac{x}{y})$ is a multiplicative c-class function [Clement Ampadu and Arslan Hojat Ansari, FIXED POINT THEOREMS IN COMPLETE MULTIPLICATIVE METRIC SPACES WITH APPLICATION TO MULTIPLICATIVE ANALOGUE OF C-CLASS FUNCTIONS, JP Journal of Fixed Point Theory and Applications, August 2016, Volume 11, Issue 2,Pages 113 - 124]. Suppose that
>
> (i) $TX \subseteq IX$ and $SX \subseteq JX$
>
> (ii) I and J are dominating maps and S and T are dominated maps
>
> (iii) If for a nondecreasing sequence $\{x_n\}$ with $y_n \preceq x_n$ for all n and $y_n \to u$ implies that $u \preceq x_n$
>
> Also assume either
>
> (a) $\{S, I\}$ are compatible, S or I is continuous and $\{T, J\}$ are weakly compatible or
>
> (b) $\{T, J\}$ are compatible, T or J is continuous and $\{S, I\}$ are weakly compatible
>
> Then S, T, I, J have a common fixed point

2.5 References

(1) Ya. I. Alber and S. Guerre-Delabriere, Principles of weakly contractive maps in Hilbert spaces, in: I. Gohberg and Yu Lyubich New Results in Operator Theory and it's Applications, in: Oper. Theory Adv. Appl. Birkhauser, Basel, 1997

(2) Clement Boateng Ampadu. A COUPLED VERSION OF THE HIGHER-ORDER BANACH CONTRACTION PRINCIPLE IN MULTIPLICATIVE CONE METRIC SPACE. JP Journal of Applied Mathematics, To Appear

(3) Ampadu, Clement (2016). Fixed Point Theory for Higher-Order Mappings. lulu.com. ISBN:5800118959925

(4) L.-G. Huang and X. Zhang, Cone metric spaces and fixed point theorems of contractive mappings, J.Math.Anal.Appl. 332 (2007) 1468-1476

(5) D. Ilic, V. Rakocevic, Quasi contraction on a cone metric space, Appl. Math. Lett. 22 (2009) 728-731

(6) G. Jungck, S. Radenovic, S. Radojevic, V. Rakocevic, Common fixed point theorems for weakly compatible pairs on cone metric spaces, Fixed Point Theory Appl. (2009). Art. ID 643849, 13 pp.

(7) J. Jachymski, Order-Theoretic Aspects of Metric Fixed Point Theory, Handbook of Metric Fixed Point Theory, Kluwer Academic Publishers, Dordrecht, 2001

(8) W. SINTUNAVARAT AND P. KUMAM, COMMON FIXED POINTS OF f-WEAK CONTRACTIONS IN CONE METRIC SPACES, Bulletin of the Iranian Mathematical Society Vol. 38 No. 2 (2012), pp 293-303

(9) Clement Ampadu and Arslan Hojat Ansari, FIXED POINT THEOREMS IN COMPLETE MULTIPLICATIVE METRIC SPACES WITH APPLICATION TO MULTIPLICATIVE ANALOGUE OF C-CLASS FUNCTIONS, JP Journal of Fixed Point Theory and Applications, August 2016, Volume 11, Issue 2,Pages 113 - 124

(10) B. S. Choudhury and N. Metiya, Fixed points of weak contractions in cone metric spaces, Nonlinear Analysis 72 (2010), no. 3-4, 1589-1593

(11) C. T. AAGE AND J. N. SALUNKE, FIXED POINTS OF (ϕ, ψ)-WEAK CONTRACTIONS IN CONE METRIC SPACES, Ann. Funct. Anal. 2 (2011), no. 1, 59–71

(12) BINAYAK S. CHOUDHURY, L. KUMAR, T. SOM; AND N. METIYA, A WEAK CONTRACTION PRINCIPLE IN PARTIALLY ORDERED CONE METRIC SPACE WITH THREE CONTROL FUNCTIONS, International Journal of Analysis and Applications, Volume 6, Number 1 (2014), 18-27

(13) Hemant Kumar Nashine and Hassen Aydi, Common fixed points for generalized (ψ, ϕ)-weak contractions in ordered cone metric spaces, Applied General Topology, Volume 13, no. 2, 2012, pp. 151-166

Chapter 3

Fixed Point Theorems for Implicit Generalized Weak Multiplicative Contraction Mappings in Multiplicative Analogue of Modular Spaces

3.1 Brief Summary

Abstract C.1 1

Multiplicative modular spaces is introduced and existence of common fixed point for a generalized weak multiplicative contraction mapping defined implicitly via multiplicative c-class functions [Clement Ampadu and Arslan Hojat Ansari, FIXED POINT THEOREMS IN COMPLETE MULTIPLICATIVE METRIC SPACES WITH APPLICATION TO MULTIPLICATIVE ANALOGUE OF C-CLASS FUNCTIONS, JP Journal of Fixed Point Theory and Applications, August 2016, Volume 11, Issue 2, Pages 113 - 124] is proved.

3.2 Preliminaries

Definition C.1 1

Let X be a vector space over $\mathbb{R}$ (or $\mathbb{C}$). A functional $\rho^* : X \mapsto [1, \infty]$ will be called a multiplicative modular if for arbitrary x and y, elements of X, it satisfies the following conditions

(a) $\rho^*(x) = 1$ iff $x = 0$

(b) $\rho^*(\alpha x) = \rho^*(x)$ for all scalar α with $|\alpha| = 1$

(c) $\rho^*(\alpha x + \beta y) \leq \rho^*(x) \cdot \rho^*(y)$, whenever $\alpha, \beta \geq 0$ and $\alpha + \beta = 1$

Remark C.2 1

If in (c) of the above definition, we have $\rho^*(\alpha x + \beta y) \leq \rho^*(x)^{\alpha^s} \cdot \rho^*(y)^{\beta^s}$, for $\alpha, \beta \geq 0$, $\alpha^s + \beta^s = 1$ with an $s \in [0, 1)$, then ρ^* will be called s-convex multiplicative modular. If $s = 1$, then ρ^* will be called convex multiplicative modular

Definition C.3 1

If ρ^* is a multiplicative modular on X, then $X_{\rho^*} = \{x \in X : \rho^*(\lambda x) \to 1, \lambda \to 0\}$ will be called a multiplicative modular space

Taking inspiration from [L. Maligranda, Orlicz spaces and interpolation, Uiversidade Estadual de Campinas Campinass SP, Brasil 1989] we introduce the following

Proposition C.4 1

Let ρ^* be a multiplicative modular on X, then

(a) $\rho^*(\alpha x)$ is a nondecreasing function of $\alpha \geq 0$

(b) If ρ^* is s-convex, then $\rho^*(\alpha x)^{\alpha^{-s}}$ is a nondecreasing function of $\alpha \geq 0$

Definition C.5 1

We will say ρ^* satisfies $\triangle_2^*$-condition if $\rho^*(2x_n) \to 1$ as $n \to \infty$, whenever $\rho^*(x_n) \to 1$ as $n \to \infty$

Definition C.6 1

Let X_{ρ^*} be a multiplicative modular space

(a) We say $\{x_n\}$ in X_{ρ^*} is ρ^*-convergent to $x \in X_{\rho^*}$ if $\rho^*(x_n - x) \to 1$ as $n \to \infty$

(b) We say $\{x_n\}$ in X_{ρ^*} is ρ^*-Cauchy if $\rho^*(x_n - x_m) \to 1$ as $n, m \to \infty$

(c) A subset C of X_{ρ^*} is said to be ρ^*-closed if the ρ^*-limit of a ρ^*-convergent sequence of C always belongs to C

(d) A subset C of X_{ρ^*} is said to be ρ^*-complete if any ρ^*-Cauchy sequence in C is a ρ^*-convergent sequence and its limit is in C

(e) A subset C of X_{ρ^*} is said to be ρ^*-bounded if $\delta_{\rho^*}(C) = \sup\{\rho^*(x-y) : x, y \in C\} < \infty$

Definition C.7 1

Let X_{ρ^*} be a multiplicative modular space, where ρ^* satisfies the $\triangle_2^*$-condition. Two self-mappings T and f of X_{ρ^*} will be called ρ^*-compatible if $\rho^*(Tfx_n - fTx_n) \to 1$ as $n \to \infty$, whenever $\{x_n\}$ is a sequence in X_{ρ^*} such that $fx_n \to z$ and $Tx_n \to z$ for some point $z \in X_{\rho^*}$

3.3 Main Results

> **Theorem C.1 1**
>
> Let X_{ρ^*} be a ρ^*-complete multiplicative modular space, where ρ^* satisfies the $\triangle_2^*$-condition. Let $c, l \in \mathbb{R}^+$, $c > l$ and $T, g : X_{\rho^*} \mapsto X_{\rho^*}$ be two ρ^*-compatible mappings such that $T(X_{\rho^*}) \subseteq g(X_{\rho^*})$ and satisfy $\psi(\rho^*(c(Tx - Ty))) \leq F\left(\frac{\psi(\rho^*(l(gx-gy)))}{\phi(\rho^*(l(gx-gy)))}\right)$ for all $x, y \in X_{\rho^*}$, where $\psi, \phi : [1, \infty) \mapsto [1, \infty)$ are both continuous and monotone nondecreasing functions with $\psi(t) = \phi(t) = 1$ iff $t = 1$, and $F(x, y) := F(\frac{x}{y})$ is a multiplicative c-class function [Clement Ampadu and Arslan Hojat Ansari, FIXED POINT THEOREMS IN COMPLETE MULTIPLICATIVE METRIC SPACES WITH APPLICATION TO MULTIPLICATIVE ANALOGUE OF C-CLASS FUNCTIONS, JP Journal of Fixed Point Theory and Applications, August 2016, Volume 11, Issue 2, Pages 113 - 124]. If one of T or g is continuous, then there exists a unique common fixed point of T and g

Proof of Theorem C.1 1

Let $x \in X_{\rho^*}$ and generate inductively the sequence $\{Tx_n\}$ as follows: $Tx_n = gx_{n+1}$. First we prove that the sequence $\{\rho^*(c(Tx_n - Tx_{n-1}))\}$ converges to one. First observe by the properties of F we have that

$$\psi(\rho^*(c(Tx_n - Tx_{n-1}))) \leq F\left(\frac{\psi(\rho^*(l(gx_n - gx_{n-1})))}{\phi(\rho^*(l(gx_n - gx_{n-1})))}\right) \tag{3.1}$$
$$\leq \psi(\rho^*(l(gx_n - gx_{n-1})))$$

On the other hand by the properties of ψ and Proposition C.4 with $c > l$ we have

$$\rho^*(c(Tx_n - Tx_{n-1})) \leq \rho^*(l(gx_n - gx_{n-1}))$$
$$= \rho^*(l(Tx_{n-1} - Tx_{n-2})) \tag{3.2}$$
$$< \rho^*(c(Tx_{n-1} - Tx_{n-2}))$$

It follows that the sequence $\{\rho^*(c(Tx_n - Tx_{n-1}))\}$ is nonincreasing and bounded below. Hence there exists $r \geq 1$ such that

$$\lim_{n \to \infty} \rho^*(c(Tx_n - Tx_{n-1})) = r$$

If $r > 1$, then taking limits in (3.2) we deduce that $\lim_{n \to \infty} \rho^*(l(gx_n - gx_{n-1})) = r$. Further taking limits in (3.1), we deduce that $\psi(r) \leq F\left(\frac{\psi(r)}{\phi(r)}\right)$ which is a contradiction, thus $r = 1$, that is, $\lim_{n \to \infty} \rho^*(c(Tx_n - Tx_{n-1})) = 1$. Next we prove that the sequence $\{cTx_n\}$ is ρ^*-Cauchy. Suppose that $\{cTx_n\}$ is not ρ^*-Cauchy, then there exists $\epsilon > 1$ and subsequence $\{x_{m_k}\}, \{x_{n_k}\}$ with $m_k > n_k \geq k$ such that $\rho^*(c(Tx_{m_k} - Tx_{n_k})) \geq \epsilon$ for $k = 1, 2, 3, \cdots$ where we can assume that $\rho^*(c(Tx_{m_k - 1} - Tx_{n_k})) < \epsilon$. Let m_k be the smallest number exceeding n_k for which $\rho^*(c(Tx_{m_k} - Tx_{n_k})) \geq \epsilon$ for $k = 1, 2, 3, \cdots$, and set

$$\theta_k = \{m \in \mathbb{N} | \exists n_k \in \mathbb{N}; \rho^*(c(Tx_{m_k} - Tx_{n_k})) \geq \epsilon, m > n_k \geq k\}$$

Since $\theta_k \subset \mathbb{N}$ and clearly, $\theta_k \neq \emptyset$. By well ordering principle, the minimum element of θ_k is denoted by m_k and obviously $\rho^*(c(Tx_{m_k - 1} - Tx_{n_k})) < \epsilon$ holds. Now let $\alpha \in \mathbb{R}^+$ be such that $\frac{l}{c} + \frac{1}{\alpha} = 1$, then we have

$$\psi(\rho^*(c(Tx_{m_k} - Tx_{n_k}))) \leq F\left[\frac{\psi(\rho^*(l(gx_{m_k} - gx_{n_k})))}{\phi(\rho^*(l(gx_{m_k} - gx_{n_k})))}\right]$$
$$\leq \psi(\rho^*(l(gx_{m_k} - gx_{n_k}))) \tag{3.3}$$
$$= \psi(\rho^*(l(Tx_{m_k - 1} - Tx_{n_k - 1})))$$

and

$$\rho^*(l(Tx_{m_k - 1} - Tx_{n_k - 1})) = \rho^*(l(Tx_{m_k - 1} - Tx_{n_k} + Tx_{n_k} - Tx_{n_k - 1}))$$
$$= \rho^*(\frac{l}{c}c(Tx_{m_k - 1} - Tx_{n_k}) + \frac{1}{\alpha}\alpha l(Tx_{n_k} - Tx_{n_k - 1})) \tag{3.4}$$
$$\leq \rho^*(c(Tx_{m_k - 1} - Tx_{n_k})) \cdot \rho^*(\alpha l(Tx_{n_k} - Tx_{n_k - 1}))$$
$$\leq \epsilon \cdot \rho^*(\alpha l(Tx_{n_k} - Tx_{n_k - 1}))$$

By $\triangle_2^*$-condition and $\lim_{n \to \infty} \rho^*(c(Tx_n - Tx_{n-1})) = 1$, we deduce that $\lim_{n \to \infty} \rho^*(\alpha l(Tx_{n_k} - Tx_{n_k - 1})) = 1$, it follows that $\lim_{k \to \infty} \psi(\rho^*(\alpha l(Tx_{n_k} - Tx_{n_k - 1}))) < \psi(\epsilon)$. Consequently, it follows that $\psi(\epsilon) \leq \lim_{k \to \infty} \psi(\rho^*(\alpha l(Tx_{n_k} - Tx_{n_k - 1}))) < \psi(\epsilon)$ which is a contradiction, hence $\{cTx_n\}$ is ρ^*-Cauchy and by the $\triangle_2^*$-condition $\{Tx_n\}$ is ρ^*-Cauchy. Since X_{ρ^*} is ρ^*-complete, there exists a point $u \in X_{\rho^*}$ such that $\rho^*(Tx_n - u) \to 1$ as $n \to \infty$, that is, $Tx_n \to u$ implies that $gx_n \to u$ as $n \to \infty$. If T is continuous, then $T^2 x_n \to Tu$ and $Tgx_n \to Tu$ as $n \to \infty$. By ρ^*-compatibility, $\rho^*(c(gTx_n - Tgx_n)) \to 1$ as $n \to \infty$, thus $gTx_n \to Tu$ as $n \to \infty$. Now we prove that u is a unique fixed point of T.

> **Proof of Theorem C.1 continued 1**
>
> Observe that
>
> $$\psi(\rho^*(c(T^2x_n - Tx_n))) = \psi(\rho^*(c(T(Tx_n) - Tx_n))) \qquad (3.5)$$
> $$\leq F\left[\begin{array}{c}\psi(\rho^*(l(gTx_n - gx_n)))\\ \phi(\rho^*(l(gTx_n - gx_n)))\end{array}\right]$$
>
> Taking limits in the inequality immediately above and using properties of F, we have,
>
> $$\psi(\rho^*(c(Tu - u))) \leq F\left[\begin{array}{c}\psi(\rho^*(l(Tu - u)))\\ \phi(\rho^*(l(Tu - u)))\end{array}\right] \qquad (3.6)$$
> $$\leq \psi(\rho^*(l(Tu - u)))$$
>
> By properties of ψ and Proposition C.4 with $c > l$ we have $\rho^*(c(Tu - u)) = 1$ and $Tu = u$. Since $T(X_{\rho^*}) \subset g(X_{\rho^*})$, then there exists a point u_1 such that $u = Tu = gu_1$. Now taking limits in the inequality
>
> $$\psi(\rho^*(c(T^2x_n - Tu_1))) = \psi(\rho^*(c(T(Tx_n) - Tx_n))) \leq F\left[\begin{array}{c}\psi(\rho^*(l(gTx_n - gu_1)))\\ \phi(\rho^*(l(gTx_n - gu_1)))\end{array}\right] \qquad (3.7)$$
>
> we deduce, using the properties of F, that
>
> $$\psi(\rho^*(c(u - Tu_1))) \leq F\left[\begin{array}{c}\psi(\rho^*(l(u - gu_1)))\\ \phi(\rho^*(l(u - gu_1)))\end{array}\right]$$
> $$\leq \psi(\rho^*(l(u - gu_1))) \qquad (3.8)$$
> $$= \psi(\rho^*(l(u - u)))$$
> $$= 1$$
>
> It follows that $u = Tu_1 = gu_1$ and also $gu = gTu_1 = Tgu_1 = Tu = u$. If g is continuous, then by a similar argument, we can show $gu = Tu = u$. Finally suppose there exists $v \in X_{\rho^*}$ such that $Tv = v = gv$ and $v \neq u$, then using properties of F, we have,
>
> $$\psi(\rho^*(c(u - v))) = \psi(\rho^*(c(Tu - Tv)))$$
> $$\leq F\left[\begin{array}{c}\psi(\rho^*(l(gu - gv)))\\ \phi(\rho^*(l(gu - gv)))\end{array}\right] \qquad (3.9)$$
> $$\leq \psi(\rho^*(l(u - v)))$$
>
> By properties of ψ and Proposition C.4 with $c > l$, we deduce that
>
> $$\rho^*(c(u - v)) < \rho^*(l(u - v)) < \rho^*(c(u - v))$$
>
> which is a contradiction. Hence $u = v$

If $\psi(t) = t$ in the above theorem, then we get the following

> **Corollary C.2 1**
>
> Let X_{ρ^*} be a ρ^*-complete multiplicative modular space, where ρ^* satisfies the $\triangle_2^*$-condition. Let $c, l \in \mathbb{R}^+$, $c > l$ and $T, g : X_{\rho^*} \mapsto X_{\rho^*}$ be two ρ^*-compatible mappings such that $T(X_{\rho^*}) \subseteq g(X_{\rho^*})$ and satisfy $\rho^*(c(Tx - Ty)) \leq F\left(\frac{\rho^*(l(gx-gy))}{\phi(\rho^*(l(gx-gy)))}\right)$ for all $x, y \in X_{\rho^*}$, where $\psi, \phi : [1, \infty) \mapsto [1, \infty)$ are both continuous and monotone nondecreasing functions with $\psi(t) = \phi(t) = 1$ iff $t = 1$, and $F(x, y) := F(\frac{x}{y})$ is a multiplicative c-class function [Clement Ampadu and Arslan Hojat Ansari, FIXED POINT THEOREMS IN COMPLETE MULTIPLICATIVE METRIC SPACES WITH APPLICATION TO MULTIPLICATIVE ANALOGUE OF C-CLASS FUNCTIONS, JP Journal of Fixed Point Theory and Applications, August 2016, Volume 11, Issue 2, Pages 113 - 124]. If one of T or g is continuous, then there exists a unique common fixed point of T and g

3.4 Exercises

Exercise C.1 1

Taking inspiration from [Chirasak Mongkolkeha and Poom Kumam, COMMON FIXED POINTS FOR GENERALIZED WEAK CONTRACTION MAPPINGS IN MODULAR SPACES, Scientiae. Mathematicae. Japonicae, Online, e-2012, 117–127] prove the following: Let X_{ρ^*} be a ρ^*-complete multiplicative modular space and let $T, S : X_{\rho^*} \mapsto X_{\rho^*}$ be mappings satisfying the inequality

$$\phi(\rho^*(Tx - Sy)) \leq \frac{\phi(M(x,y))}{\psi(M(x,y))}$$

for all $x, y \in X_{\rho^*}$, where $M(x,y) = \{\rho^*(x - y), \rho^*(x - Tx), \rho^*(y - Sy), \sqrt{\rho^*(\frac{1}{2}(y - Tx)) \cdot \rho^*(\frac{1}{2}(x - Sy))}\}$ and $\phi, \psi : [1, \infty) \mapsto [1, \infty)$ are both continuous and monotone nondecreasing functions with $\phi(t) = \psi(t) = 1$ iff $t = 1$. Then there exists a unique point $u \in X_{\rho^*}$ such that $u = Tu = Su$

Exercise C.2 1

Taking inspiration from Theorem 12 [Chirasak Mongkolkeha and Poom Kumam, Some Fixed Point Results for Generalized Weak Contraction Mappings in Modular Spaces, International Journal of Analysis Volume 2013, Article ID 247378, 6 pages] deduce that Exercise C.1 holds if $T = S$

Exercise C.3 1

Taking inspiration from Theorem 10 [Chirasak Mongkolkeha and Poom Kumam, Some Fixed Point Results for Generalized Weak Contraction Mappings in Modular Spaces, International Journal of Analysis Volume 2013, Article ID 247378, 6 pages] deduce that Theorem C.1 holds if g is the identity

Exercise C.4 1

Give the statements in the following cases:

(a) Corollary arising from Theorem C.1 when ψ and g are both the identity

(b) Corollary arising from Exercise C.1 when ϕ is the identity and $T = S$

3.5 References

(1) Clement Ampadu and Arslan Hojat Ansari, FIXED POINT THEOREMS IN COMPLETE MULTIPLICATIVE METRIC SPACES WITH APPLICATION TO MULTIPLICATIVE ANALOGUE OF C-CLASS FUNCTIONS, JP Journal of Fixed Point Theory and Applications, August 2016, Volume 11, Issue 2, Pages 113 - 124

(2) L. Maligranda, Orlicz spaces and interpolation, Uiversidade Estadual de Campinas Campinass SP, Brasil 1989

(3) Chirasak Mongkolkeha and Poom Kumam, COMMON FIXED POINTS FOR GENERALIZED WEAK CONTRACTION MAPPINGS IN MODULAR SPACES, Scientiae. Mathematicae. Japonicae, Online, e-2012, 117–127

(4) Chirasak Mongkolkeha and Poom Kumam, Some Fixed Point Results for Generalized Weak Contraction Mappings in Modular Spaces, International Journal of Analysis Volume 2013, Article ID 247378, 6 pages

Chapter 4

Fixed Point Theorems for Implicit Weakly Multiplicative Contractions of the Derivative Type in Multiplicative Analogue of T_0-Quasi-Metric Spaces

4.1 Brief Summary

Abstract D.1 1

In [Y. U. Gaba, Unique fixed point theorems for contractive maps type in T_0-quasi-metric spaces, Adv. Fixed Point Theory, 4 (2014), 117-124] fixed point results for C-contractive and S-contractive self mappings defined in T_0-quasi metric spaces was proved, and in [ENIOLA FUNMILAYO KAZEEM, YAE ULRICH GABA, WEAKLY CONTRACTIVE MAPPINGS IN T_0-QUASI-METRIC SPACES, Adv. Fixed Point Theory, 4 (2014), No. 3, 355-364] existence of a fixed point for weakly C-contractive and weakly S-contractive self mappings defined in T_0-quasi metric spaces was proved. Motivated by certain results obtained in [Clement Ampadu, Arslan Hojat Ansari and Memudu Olaposi Olatinwo, FIXED POINT THEOREMS USING MULTIPLICATIVE CONTRACTIVE DEFINITIONS WITH APPLICATION TO MULTIPLICATIVE ANALOGUE OF C-CLASS FUNCTIONS, JP Journal of Fixed Point Theory and Applications, To Appear] this chapter proves some fixed point theorems in the multiplicative analogue of T_0-Quasi-Metric Spaces.

4.2 Preliminaries

Definition D.1 1

Let X be a nonempty set. A function $m : X \times X \mapsto [1, \infty)$ will be called a multiplicative quasi-pseudometric on X iff

  (a) $m(x, x) = 1$ for all $x \in X$

  (b) $m(x, z) \leq m(x, y) \cdot m(y, z)$ for all $x, y, z \in X$

Moreover, if $m(x, y) = 1 = m(y, x)$, then m will be called a multiplicative T_0-quasi-pseudometric or a multiplicative di-metric

CHAPTER 4. FIXED POINT THEOREMS FOR IMPLICIT WEAKLY MULTIPLICATIVE CONTRACTIONS OF THE DERIVATIVE TYPE IN MULTIPLICATIVE ANALOGUE OF T_0-QUASI-METRIC SPACES

From [E. Kemajou, H.-P. A. Kunzi, O. O. Otafudu, The Isbell-hull of a di-space, Topology Appl. 159 (2012), 2463-2475] we deduce the following

> **Example D.2 1**
>
> Define $m : \mathbb{R} \times \mathbb{R} \mapsto [1, \infty)$ by $m(a,b) = \frac{a}{b} = \max\{\frac{a}{b}, 1\}$, then $(\mathbb{R}, m)$ is a multiplicative di-metric space

> **Remark D.3 1**
>
> Let m be a multiplicative quasi-pseudometric on X, then the map m^{-1} defined by $m^{-1}(x,y) = m(y,x)$ whenever $x, y \in X$ is also a multiplicative quasi-pseudometric on X, which we will call the conjugate of m

> **Remark D.4 1**
>
> Define $m^s := m \vee m^{-1}$ by $m^s(x,y) = \max\{m(x,y), m(y,x)\}$, then m^s is a multiplicative metric on X whenever m is a multiplicative T_0-quasi pseudo-metric

> **Definition D.5 1**
>
> The multiplicative di-metric space (X, m) will be called multiplicative bicomplete if the multiplicative metric space (X, m^s) is multiplicative complete

> **Example D.6 1**
>
> Let $X = [1, \infty)$. Define for each $x, y \in X$, $m(x,y) = x$ if $x > y$ and $m(x,y) = 1$ if $x \leq y$, then (X, m) is a multiplicative T_0-quasi-pseudo-metric space. Further for $x, y \in X$, define m^s by $m^s(x,y) = \max\{x, y\}$ if $x \neq y$ and $m^s(x,y) = 1$ if $x = y$, then m^s is a multiplicative metric on X

> **Definition D.7 1**
>
> Let (X, m) be a multiplicative quasi-pseudometric space. For each $x \in X$ and $\epsilon > 1$
>
> $$B_m(x, \epsilon) = \{y \in X : m(x,y) < \epsilon\}$$
>
> will denote the multiplicative open ϵ-ball at x. The collection of all such balls is a base for the topology $\tau(m)$ induced by m on X. Similarly, for $x \in X$ and $\epsilon > 1$
>
> $$C_m(x, \epsilon) = \{y \in X : m(x,y) \leq \epsilon\}$$
>
> will denote the multiplicative closed ϵ-ball at x

> **Definition D.8 1**
>
> [ENIOLA FUNMILAYO KAZEEM, YAE ULRICH GABA, WEAKLY CONTRACTIVE MAPPINGS IN T_0-QUASI-METRIC SPACES, Adv. Fixed Point Theory, 4 (2014), No. 3, 355-364] Let $E_1, E_2, \cdots, E_n, K$ be totally ordered spaces with respective orders $\leq_{E_1}, \leq_{E_2}, \cdots, \leq_{E_n}, \leq_{E_k}$. A map $f : E_1 \times E_2 \times \cdots \times E_n \mapsto K$ will be called component non-increasing if
>
> $$f(x_1, \cdots, x_n) \leq_K f(a_1, \cdots, a_n)$$
>
> whenever $a_i \leq_{E_i} x_i$ for any $i = 1, \cdots, n$

Example D.9 1

Let $E_1 = E_2 = \cdots = E_n = K = [1, \infty)$ with natural ordering and define $f : [1, \infty) \times \cdots \times [1, \infty) \mapsto [1, \infty)$ by $f(x_1, x_2, \cdots, x_n) = \frac{1}{\sum_{i=1}^{n} x_i^2}$, then f is component non-increasing

Definition D.10 1

[Clement Ampadu, Arslan Hojat Ansari and Memudu Olaposi Olatinwo, FIXED POINT THEOREMS USING MULTIPLICATIVE CONTRACTIVE DEFINITIONS WITH APPLICATION TO MULTIPLICATIVE ANALOGUE OF C-CLASS FUNCTIONS, JP Journal of Fixed Point Theory and Applications, To Appear] Let (X, m) be a multiplicative metric space, and let $f : X \mapsto X$. We say that $f : X \mapsto X$ is a weak C-multiplicative contraction of the derivative type if

$$\frac{d\varphi}{dt}\big|_{t=m(fx,fy)} \leq \frac{\frac{d\varphi}{dt}\big|_{t=\sqrt{m(x,fy) \cdot m(y,fx)}}}{\frac{d\varphi}{dt}\big|_{t=\psi(m(x,fy),m(y,fx))}}$$

where $\psi : [1, \infty)^2 \mapsto [1, \infty)$ is a continuous mapping such that $\psi(x, y) = 1$ iff $x = y = 1$ and $\varphi : [1, \infty) \mapsto [1, \infty)$ is such that $\frac{d\varphi}{dt}\big|_{t=\epsilon} > 1$ for each $\epsilon > 1$

By using the multiplicative c-class function [Clement Ampadu and Arslan Hojat Ansari,FIXED POINT THEOREMS IN COMPLETE MULTIPLICATIVE METRIC SPACES WITH APPLICATION TO MULTIPLICATIVE ANALOGUE OF C-CLASS FUNCTIONS, JP Journal of Fixed Point Theory and Applications, August 2016, Volume 11, Issue 2,Pages 113 - 124] the above definition can be written implicitly as follows

Definition D.11 1

Let (X, m) be a multiplicative metric space, and let $f : X \mapsto X$. We say that $f : X \mapsto X$ is an implicit weak C-multiplicative contraction of the derivative type if

$$\frac{d\varphi}{dt}\big|_{t=m(fx,fy)} \leq F\left[\frac{\frac{d\varphi}{dt}\big|_{t=\sqrt{m(x,fy) \cdot m(y,fx)}}}{\frac{d\varphi}{dt}\big|_{t=\psi(m(x,fy),m(y,fx))}}\right]$$

where $\psi : [1, \infty)^2 \mapsto [1, \infty)$ is a continuous mapping such that $\psi(x, y) = 1$ iff $x = y = 1$, $\varphi : [1, \infty) \mapsto [1, \infty)$ is such that $\frac{d\varphi}{dt}\big|_{t=\epsilon} > 1$ for each $\epsilon > 1$, and $F(x, y) := F(\frac{x}{y})$ is a multiplicative c-class function [Clement Ampadu and Arslan Hojat Ansari,FIXED POINT THEOREMS IN COMPLETE MULTIPLICATIVE METRIC SPACES WITH APPLICATION TO MULTIPLICATIVE ANALOGUE OF C-CLASS FUNCTIONS, JP Journal of Fixed Point Theory and Applications, August 2016, Volume 11, Issue 2,Pages 113 - 124]

Definition D.12 1

[Clement Ampadu, Arslan Hojat Ansari and Memudu Olaposi Olatinwo, FIXED POINT THEOREMS USING MULTIPLICATIVE CONTRACTIVE DEFINITIONS WITH APPLICATION TO MULTIPLICATIVE ANALOGUE OF C-CLASS FUNCTIONS, JP Journal of Fixed Point Theory and Applications, To Appear] Let (X, m) be a multiplicative metric space, and let $f : X \mapsto X$. We say that $f : X \mapsto X$ is a weak S-multiplicative contraction of the derivative type if

$$\frac{d\varphi}{dt}\big|_{t=m(fx,fy)} \leq \frac{\sqrt[3]{\frac{d\varphi}{dt}\big|_{t=m(x,fy)} \cdot \frac{d\varphi}{dt}\big|_{t=m(y,fx)} \cdot \frac{d\varphi}{dt}\big|_{t=m(x,y)}}}{\psi\left(\frac{d\varphi}{dt}\big|_{t=m(x,fy)}, \frac{d\varphi}{dt}\big|_{t=m(y,fx)}, \frac{d\varphi}{dt}\big|_{t=m(x,y)}\right)}$$

where $\psi : [1, \infty)^3 \mapsto [1, \infty)$ is a continuous mapping such that $\psi(x, y, z) = 1$ iff $x = y = z = 1$ and $\varphi : [1, \infty) \mapsto [1, \infty)$ is such that $\frac{d\varphi}{dt}\big|_{t=\epsilon} > 1$ for each $\epsilon > 1$

By using the multiplicative c-class function [Clement Ampadu and Arslan Hojat Ansari,FIXED POINT THEOREMS IN COMPLETE MULTIPLICATIVE METRIC SPACES WITH APPLICATION TO MULTIPLICATIVE ANALOGUE OF C-CLASS FUNCTIONS, JP Journal of Fixed Point Theory and Applications, August 2016, Volume 11, Issue 2,Pages 113 - 124] the above definition can be written implicitly as follows

> **Definition D.13 1**
>
> Let (X, m) be a multiplicative metric space, and let $f : X \mapsto X$. We say that $f : X \mapsto X$ is an implicit weak S-multiplicative contraction of the derivative type if
>
> $$\frac{d\varphi}{dt}|_{t=m(fx,fy)} \leq F\left[\frac{\sqrt[3]{\frac{d\varphi}{dt}|_{t=m(x,fy)} \cdot \frac{d\varphi}{dt}|_{t=m(y,fx)} \cdot \frac{d\varphi}{dt}|_{t=m(x,y)}}}{\psi(\frac{d\varphi}{dt}|_{t=m(x,fy)}, \frac{d\varphi}{dt}|_{t=m(y,fx)}, \frac{d\varphi}{dt}|_{t=m(x,y)})}\right]$$
>
> where $\psi : [1, \infty)^3 \mapsto [1, \infty)$ is a continuous mapping such that $\psi(x, y, z) = 1$ iff $x = y = z = 1$ and $\varphi : [1, \infty) \mapsto [1, \infty)$ is such that $\frac{d\varphi}{dt}|_{t=\epsilon} > 1$ for each $\epsilon > 1$, and $F(x, y) := F(\frac{x}{y})$ is a multiplicative c-class function [Clement Ampadu and Arslan Hojat Ansari,FIXED POINT THEOREMS IN COMPLETE MULTIPLICATIVE METRIC SPACES WITH APPLICATION TO MULTIPLICATIVE ANALOGUE OF C-CLASS FUNCTIONS, JP Journal of Fixed Point Theory and Applications, August 2016, Volume 11, Issue 2,Pages 113 - 124]

> **Theorem D.14 1**
>
> [Clement Ampadu, Arslan Hojat Ansari and Memudu Olaposi Olatinwo, FIXED POINT THEOREMS USING MULTIPLICATIVE CONTRACTIVE DEFINITIONS WITH APPLICATION TO MULTIPLICATIVE ANALOGUE OF C-CLASS FUNCTIONS, JP Journal of Fixed Point Theory and Applications, To Appear] Let (X, m) be a complete multiplicative metric space. If $f : X \mapsto X$ is a weak C-multiplicative contraction of the derivative type, then $f : X \mapsto X$ has a unique fixed point

Using the implicit weak C-multiplicative contraction of the derivative type (Definition D.11), the above theorem can be generalized as follows

> **Theorem D.15 1**
>
> Let (X, m) be a complete multiplicative metric space. If $f : X \mapsto X$ is an implicit weak C-multiplicative contraction of the derivative type, then $f : X \mapsto X$ has a unique fixed point

> **Theorem D.16 1**
>
> [Clement Ampadu, Arslan Hojat Ansari and Memudu Olaposi Olatinwo, FIXED POINT THEOREMS USING MULTIPLICATIVE CONTRACTIVE DEFINITIONS WITH APPLICATION TO MULTIPLICATIVE ANALOGUE OF C-CLASS FUNCTIONS, JP Journal of Fixed Point Theory and Applications, To Appear] Let (X, m) be a complete multiplicative metric space. If $f : X \mapsto X$ is a weak S-multiplicative contraction of the derivative type, then $f : X \mapsto X$ has a unique fixed point

Using the implicit weak S-multiplicative contraction of the derivative type (Definition D.13), the above theorem can be generalized as follows

> **Theorem D.17 1**
>
> Let (X, m) be a complete multiplicative metric space. If $f : X \mapsto X$ is an implicit weak S-multiplicative contraction of the derivative type, then $f : X \mapsto X$ has a unique fixed point

> **Definition D.18 1**
>
> Let (X, m) be a multiplicative quasi-pseudo-metric space.
>
> (a) A map $f : X \mapsto X$ satisfying the inequality in Definition D.10 will be called a weak C-multiplicative pseudo-contraction of the derivative type
>
> (b) A map $f : X \mapsto X$ satisfying the inequality in Definition D.11 will be called an implicit weak C-multiplicative pseudo-contraction of the derivative type
>
> (c) A map $f : X \mapsto X$ satisfying the inequality in Definition D.12 will be called a weak S-multiplicative pseudo-contraction of the derivative type
>
> (d) A map $f : X \mapsto X$ satisfying the inequality in Definition D.13 will be called an implicit weak S-multiplicative pseudo-contraction of the derivative type

4.3 Main Results

> **Theorem D.1 1**
>
> Let (X, m) be a totally ordered multiplicative bi-complete multiplicative di-metric space, and let $f : X \mapsto X$ be an implicit weak C-multiplicative pseudo-contraction of the derivative type. Moreover, assume that ψ is component non-increasing. Then T has a unique fixed point.

Proof of Theorem D.1 1

Since $f : X \mapsto X$ is an implicit weak C-multiplicative pseudo-contraction of the derivative type, it follows that the inequality in Definition D.11 holds. Now for any $x, y \in X$ we have

$$\frac{d\varphi}{dt}\Big|_{t=m^{-1}(fx,fy)} = \frac{d\varphi}{dt}\Big|_{t=m(fy,fx)}$$

$$\leq F\begin{bmatrix} \frac{d\varphi}{dt}\big|_{t=\sqrt{m(y,fx)\cdot m(x,fy)}} \\ \frac{d\varphi}{dt}\big|_{t=\psi(m(y,fx),m(x,fy))} \end{bmatrix}$$

$$\leq F\begin{bmatrix} \frac{d\varphi}{dt}\big|_{t=\sqrt{m^{-1}(fx,y)\cdot m^{-1}(fy,x)}} \\ \frac{d\varphi}{dt}\big|_{t=\psi(m^{-1}(fx,y),m^{-1}(fy,x))} \end{bmatrix}$$

It follows that $f : (X, m^{-1}) \mapsto (X, m^{-1})$ is an implicit weak C-multiplicative pseduo-contraction of the derivative type. Now since ψ is component non-increasing we have,

$$\frac{d\varphi}{dt}\Big|_{t=m(fx,fy)} \leq F\begin{bmatrix} \frac{d\varphi}{dt}\big|_{t=\sqrt{m(x,fy)\cdot m(y,fx)}} \\ \frac{d\varphi}{dt}\big|_{t=\psi(m(x,fy),m(y,fx))} \end{bmatrix}$$

$$\leq F\begin{bmatrix} \frac{d\varphi}{dt}\big|_{t=\sqrt{m^s(x,fy)\cdot m^s(y,fx)}} \\ \frac{d\varphi}{dt}\big|_{t=\psi(m^s(x,fy),m^s(y,fx))} \end{bmatrix}$$

and

$$\frac{d\varphi}{dt}\Big|_{t=m^{-1}(fx,fy)} \leq F\begin{bmatrix} \frac{d\varphi}{dt}\big|_{t=\sqrt{m^{-1}(fx,y)\cdot m^{-1}(fy,x)}} \\ \frac{d\varphi}{dt}\big|_{t=\psi(m^{-1}(fx,y),m^{-1}(fy,x))} \end{bmatrix}$$

$$\leq F\begin{bmatrix} \frac{d\varphi}{dt}\big|_{t=\sqrt{m^s(x,fy)\cdot m^s(y,fx)}} \\ \frac{d\varphi}{dt}\big|_{t=\psi(m^s(x,fy),m^s(y,fx))} \end{bmatrix}$$

for all $x, y \in X$. Hence,

$$\frac{d\varphi}{dt}\Big|_{t=m^s(fx,fy)} \leq F\begin{bmatrix} \frac{d\varphi}{dt}\big|_{t=\sqrt{m^s(x,fy)\cdot m^s(y,fx)}} \\ \frac{d\varphi}{dt}\big|_{t=\psi(m^s(x,fy),m^s(y,fx))} \end{bmatrix}$$

for all $x, y \in X$. Thus, $f : (X, m^s) \mapsto (X, m^s)$ is an implicit weak C-multiplicative contraction of the derivative type. By assumption (X, m) is multiplicative bicomplete, hence (X, m^s) is multiplicative complete. Therefore by Theorem D.15, T has a unique fixed point

Theorem D.2 1

Let (X, m) be a totally ordered multiplicative bi-complete multiplicative di-metric space, and let $f : X \mapsto X$ be an implicit weak S-multiplicative pseudo-contraction of the derivative type. Moreover, assume that ψ is component non-increasing. Then T has a unique fixed point.

Proof of Theorem D.2 1

Since $f : X \mapsto X$ is an implicit weak S-multiplicative pseudo-contraction of the derivative type, it follows that the inequality in Definition D.13 holds. Now for any $x, y \in X$ we have

$$\frac{d\varphi}{dt}\Big|_{t=m^{-1}(fx,fy)} = \frac{d\varphi}{dt}\Big|_{t=m(fy,fx)}$$

$$\leq F\left[\frac{\sqrt[3]{\frac{d\varphi}{dt}\big|_{t=m(y,fx)} \cdot \frac{d\varphi}{dt}\big|_{t=m(x,fy)} \cdot \frac{d\varphi}{dt}\big|_{t=m(y,x)}}}{\psi\left(\frac{d\varphi}{dt}\big|_{t=m(y,fx)}, \frac{d\varphi}{dt}\big|_{t=m(x,fy)}, \frac{d\varphi}{dt}\big|_{t=m(y,x)}\right)}\right]$$

$$\leq F\left[\frac{\sqrt[3]{\frac{d\varphi}{dt}\big|_{t=m^{-1}(fx,y)} \cdot \frac{d\varphi}{dt}\big|_{t=m^{-1}(fy,x)} \cdot \frac{d\varphi}{dt}\big|_{t=m^{-1}(x,y)}}}{\psi\left(\frac{d\varphi}{dt}\big|_{t=m^{-1}(fx,y)}, \frac{d\varphi}{dt}\big|_{t=m^{-1}(fy,x)}, \frac{d\varphi}{dt}\big|_{t=m^{-1}(x,y)}\right)}\right]$$

It follows that $f : (X, m^{-1}) \mapsto (X, m^{-1})$ is an implicit weak S-multiplicative pseduo-contraction of the derivative type. Now since ψ is component non-increasing we have,

$$\frac{d\varphi}{dt}\Big|_{t=m(fx,fy)} \leq F\left[\frac{\sqrt[3]{\frac{d\varphi}{dt}\big|_{t=m(x,fy)} \cdot \frac{d\varphi}{dt}\big|_{t=m(y,fx)} \cdot \frac{d\varphi}{dt}\big|_{t=m(x,y)}}}{\psi\left(\frac{d\varphi}{dt}\big|_{t=m(x,fy)}, \frac{d\varphi}{dt}\big|_{t=m(y,fx)}, \frac{d\varphi}{dt}\big|_{t=m(x,y)}\right)}\right]$$

$$\leq F\left[\frac{\sqrt[3]{\frac{d\varphi}{dt}\big|_{t=m^s(x,fy)} \cdot \frac{d\varphi}{dt}\big|_{t=m^s(y,fx)} \cdot \frac{d\varphi}{dt}\big|_{t=m^s(x,y)}}}{\psi\left(\frac{d\varphi}{dt}\big|_{t=m^s(x,fy)}, \frac{d\varphi}{dt}\big|_{t=m^s(y,fx)}, \frac{d\varphi}{dt}\big|_{t=m^s(x,y)}\right)}\right]$$

and

$$\frac{d\varphi}{dt}\Big|_{t=m^{-1}(fx,fy)} \leq F\left[\frac{\sqrt[3]{\frac{d\varphi}{dt}\big|_{t=m^{-1}(fx,y)} \cdot \frac{d\varphi}{dt}\big|_{t=m^{-1}(fy,x)} \cdot \frac{d\varphi}{dt}\big|_{t=m^{-1}(x,y)}}}{\psi\left(\frac{d\varphi}{dt}\big|_{t=m^{-1}(fx,y)}, \frac{d\varphi}{dt}\big|_{t=m^{-1}(fy,x)}, \frac{d\varphi}{dt}\big|_{t=m^{-1}(x,y)}\right)}\right]$$

$$\leq F\left[\frac{\sqrt[3]{\frac{d\varphi}{dt}\big|_{t=m^s(x,fy)} \cdot \frac{d\varphi}{dt}\big|_{t=m^s(y,fx)} \cdot \frac{d\varphi}{dt}\big|_{t=m^s(x,y)}}}{\psi\left(\frac{d\varphi}{dt}\big|_{t=m^s(x,fy)}, \frac{d\varphi}{dt}\big|_{t=m^s(y,fx)}, \frac{d\varphi}{dt}\big|_{t=m^s(x,y)}\right)}\right]$$

for all $x, y \in X$. Hence,

$$\frac{d\varphi}{dt}\Big|_{t=m^s(fx,fy)} \leq F\left[\frac{\sqrt[3]{\frac{d\varphi}{dt}\big|_{t=m^s(x,fy)} \cdot \frac{d\varphi}{dt}\big|_{t=m^s(y,fx)} \cdot \frac{d\varphi}{dt}\big|_{t=m^s(x,y)}}}{\psi\left(\frac{d\varphi}{dt}\big|_{t=m^s(x,fy)}, \frac{d\varphi}{dt}\big|_{t=m^s(y,fx)}, \frac{d\varphi}{dt}\big|_{t=m^s(x,y)}\right)}\right]$$

for all $x, y \in X$. Thus, $f : (X, m^s) \mapsto (X, m^s)$ is an implicit weak S-multiplicative contraction of the derivative type. By assumption (X, m) is multiplicative bicomplete, hence (X, m^s) is multiplicative complete. Therefore by Theorem D.17, T has a unique fixed point.

By taking the multiplicative c-class function to be $F(x, y) := F\left(\frac{x}{y}\right) = \frac{x}{y}$, then the following are immediate

Corollary D.3 1

Let (X, m) be a totally ordered multiplicative bi-complete multiplicative di-metric space, and let $f : X \mapsto X$ be a weak C-multiplicative pseudo-contraction of the derivative type. Moreover, assume that ψ is component non-increasing. Then T has a unique fixed point.

Corollary D.4 1

Let (X, m) be a totally ordered multiplicative bi-complete multiplicative di-metric space, and let $f : X \mapsto X$ be a weak S-multiplicative pseudo-contraction of the derivative type. Moreover, assume that ψ is component non-increasing. Then T has a unique fixed point.

4.4 Exercises

Exercise D.1 1

Taking inspiration from [Vahid Parvaneh, SOME COMMON FIXED POINT THEOREMS IN COMPLETE METRIC SPACES, International Journal of Pure and Applied Mathematics, Volume 76 No. 1 2012, 1-8] prove the following: Let (X, m) be a complete multiplicative metric space and let E be a nonempty closed subset of X. Let $T, S : E \mapsto E$ be such that

$$\frac{d\varphi}{dt}|_{t=m(Tx,Sy)} \leq F\left[\frac{\frac{d\varphi}{dt}|_{t=\sqrt{m(Rx,Sy) \cdot m(Ry,Tx)}}}{\frac{d\varphi}{dt}|_{t=\psi(m(Rx,Sy),m(Ry,Tx))}}\right]$$

where $\psi : [1, \infty)^2 \mapsto [1, \infty)$ is a continuous mapping such that $\psi(x, y) = 1$ iff $x = y = 1$, $\varphi : [1, \infty) \mapsto [1, \infty)$ is such that $\frac{d\varphi}{dt}|_{t=\epsilon} > 1$ for each $\epsilon > 1$, and $F(x, y) := F(\frac{x}{y})$ is a multiplicative c-class function [Clement Ampadu and Arslan Hojat Ansari, FIXED POINT THEOREMS IN COMPLETE MULTIPLICATIVE METRIC SPACES WITH APPLICATION TO MULTIPLICATIVE ANALOGUE OF C-CLASS FUNCTIONS, JP Journal of Fixed Point Theory and Applications, August 2016, Volume 11, Issue 2, Pages 113 - 124]. Further let $R : E \mapsto X$ satisfy the following

(a) $TE \subseteq RE$ and $SE \subseteq RE$

(b) the pairs (T, R) and (S, R) are weakly compatible

In addition assume that RE is a closed subset of X. Then T, R, S have a unique common fixed point

Exercise D.2 1

Taking inspiration from [Vahid Parvaneh, SOME COMMON FIXED POINT THEOREMS IN COMPLETE METRIC SPACES, International Journal of Pure and Applied Mathematics, Volume 76 No. 1 2012, 1-8] prove the following: Let (X, m) be a complete multiplicative metric space and let E be a nonempty closed subset of X. Let $T, S : E \mapsto E$ be such that

$$\frac{d\varphi}{dt}|_{t=m(Tx,Sy)} \leq F\left[\frac{\sqrt[3]{\frac{d\varphi}{dt}|_{t=m(Rx,Sy)} \cdot \frac{d\varphi}{dt}|_{t=m(Ry,Tx)} \cdot \frac{d\varphi}{dt}|_{t=m(Rx,Ry)}}}{\psi(\frac{d\varphi}{dt}|_{t=m(Rx,Sy)}, \frac{d\varphi}{dt}|_{t=m(Ry,Tx)}, \frac{d\varphi}{dt}|_{t=m(Rx,Ry)})}\right]$$

where $\psi : [1, \infty)^3 \mapsto [1, \infty)$ is a continuous mapping such that $\psi(x, y, z) = 1$ iff $x = y = z = 1$ and $\varphi : [1, \infty) \mapsto [1, \infty)$ is such that $\frac{d\varphi}{dt}|_{t=\epsilon} > 1$ for each $\epsilon > 1$, and $F(x, y) := F(\frac{x}{y})$ is a multiplicative c-class function [Clement Ampadu and Arslan Hojat Ansari, FIXED POINT THEOREMS IN COMPLETE MULTIPLICATIVE METRIC SPACES WITH APPLICATION TO MULTIPLICATIVE ANALOGUE OF C-CLASS FUNCTIONS, JP Journal of Fixed Point Theory and Applications, August 2016, Volume 11, Issue 2, Pages 113 - 124]. Further let $R : E \mapsto X$ satisfy the following

(a) $TE \subseteq RE$ and $SE \subseteq RE$

(b) the pairs (T, R) and (S, R) are weakly compatible

In addition assume that RE is a closed subset of X. Then T, R, S have a unique common fixed point

Exercise D.3 1

Let (X, m) be a multiplicative T_0 quasi-pseudo metric space, then $m^s := m \vee m^{-1}$, that is, $m^s(x, y) = \max\{m(x, y), m(y, x)\}$ defines a multiplicative metric on X. We say that $E \subset X$ is join closed if it is m^s-closed, that is, closed with respect to the topology generated by m^s. Taking inspiration from [ENIOLA FUNMILAYO KAZEEM, YAE ULRICH GABA, WEAKLY CONTRACTIVE MAPPINGS IN T_0-QUASI-METRIC SPACES, Adv. Fixed Point Theory, 4 (2014), No. 3, 355-364] prove the following: Let $T, S : E \mapsto E$ and $R : E \mapsto X$ satisfy the inequality in Exercise D.1 and let R satisfy (a) and (b) of Exercise D.1. Assume further that RE is a join closed subset of X. Then T, R, S have a unique common fixed point

Exercise D.4 1

Let (X, m) be a multiplicative T_0 quasi-pseudo metric space, then $m^s := m \vee m^{-1}$, that is, $m^s(x, y) = \max\{m(x, y), m(y, x)\}$ defines a multiplicative metric on X. We say that $E \subset X$ is join closed if it is m^s-closed, that is, closed with respect to the topology generated by m^s. Taking inspiration from [ENIOLA FUNMILAYO KAZEEM, YAE ULRICH GABA, WEAKLY CONTRACTIVE MAPPINGS IN T_0-QUASI-METRIC SPACES, Adv. Fixed Point Theory, 4 (2014), No. 3, 355-364] prove the following: Let $T, S : E \mapsto E$ and $R : E \mapsto X$ satisfy the inequality in Exercise D.2 and let R satisfy (a) and (b) of Exercise D.2. Assume further that RE is a join closed subset of X. Then T, R, S have a unique common fixed point

4.5 References

(1) Y. U. Gaba, Unique fixed point theorems for contractive maps type in T_0-quasi-metric spaces, Adv. Fixed Point Theory, 4 (2014), 117-124

(2) ENIOLA FUNMILAYO KAZEEM, YAE ULRICH GABA, WEAKLY CONTRACTIVE MAPPINGS IN T_0-QUASI-METRIC SPACES, Adv. Fixed Point Theory, 4 (2014), No. 3, 355-364

(3) Clement Ampadu, Arslan Hojat Ansari and Memudu Olaposi Olatinwo, FIXED POINT THEOREMS USING MULTIPLICATIVE CONTRACTIVE DEFINITIONS WITH APPLICATION TO MULTIPLICATIVE ANALOGUE OF C-CLASS FUNCTIONS, JP Journal of Fixed Point Theory and Applications, To Appear

(4) E. Kemajou, H.-P. A. Kunzi, O. O. Otafudu, The Isbell-hull of a di-space, Topology Appl. 159 (2012), 2463-2475

(5) Clement Ampadu and Arslan Hojat Ansari,FIXED POINT THEOREMS IN COMPLETE MULTIPLICATIVE METRIC SPACES WITH APPLICATION TO MULTIPLICATIVE ANALOGUE OF C-CLASS FUNCTIONS, JP Journal of Fixed Point Theory and Applications, August 2016, Volume 11, Issue 2,Pages 113 - 124

(6) Vahid Parvaneh, SOME COMMON FIXED POINT THEOREMS IN COMPLETE METRIC SPACES, International Journal of Pure and Applied Mathematics, Volume 76 No. 1 2012, 1-8